Light Jagdpanzer

Light
Jagdpanzer

Development • Production • Operations

Walter J. Spielberger, Hilary L. Doyle & Thomas L. Jentz

Schiffer Military History
Atglen, PA

The sometimes poor quality of the pictures is due to their age.

The 1/24 scale drawings by Hilary J. Doyle were reduced to 1/35 scale.

Book translation by Dr. Edward Force, Central Connecticut State

Book Design by Ian Robertson.

Printed in China.
ISBN: 978-0-7643-2623-3

This book was originally published in German under the title
Leichte Jagdpanzer Entwicklung - Fertigung - Einsatz by Motorbuch Verlag

We are interested in hearing from authors with book ideas on related topics.

Published by Schiffer Publishing Ltd.
4880 Lower Valley Road
Atglen, PA 19310
Phone: (610) 593-1777
FAX: (610) 593-2002
E-mail: Info@schifferbooks.com.
Visit our web site at: www.schifferbooks.com
Please write for a free catalog.
This book may be purchased from the publisher.
Please include $3.95 postage.
Try your bookstore first.

In Europe, Schiffer books are distributed by:
Bushwood Books
6 Marksbury Avenue
Kew Gardens
Surrey TW9 4JF, England
Phone: 44 (0) 20 8392-8585
FAX: 44 (0) 20 8392-9876
E-mail: Info@bushwoodbooks.co.uk.
Visit our website at: www.bushwoodbooks.co.uk
Free postage in the UK. Europe: air mail at cost.
Try your bookstore first.

Contents

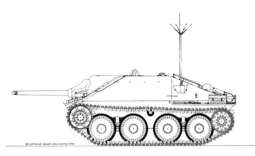

Jagdpanzer 38 - March 1944 Page 19

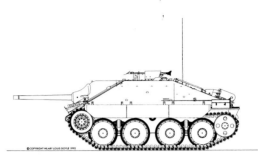

Jagdpanzer 38 - April 1945 Page 48

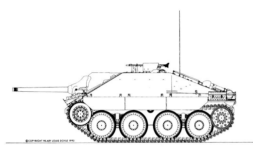

Jagdpanzer 38 - May 1944 Page 28

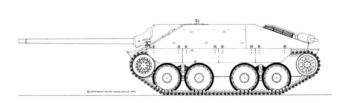

Jagdpanzer 38 D
Text, p. 52; drawings, p. 73

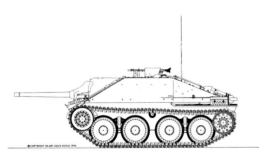

Jagdpanzer 38 - July 1944 Page 34

Jagdpanzer 38 (fixed gun)
Text, p. 52; photos and drawings, p. 70

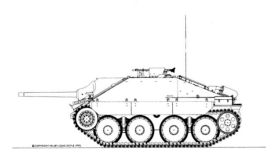

Jagdpanzer 38 - December 1944 Page 42

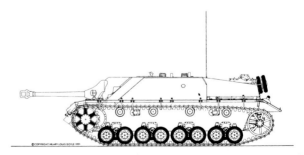

Jagdpanzer IV - 0 series Page 100

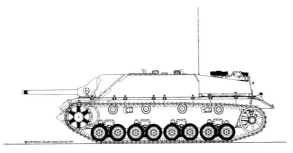

Jagdpanzer IV - April 1944 Page 121

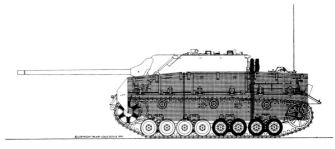

Panzer IV/70 (A) - Dec. 1944 Page 156

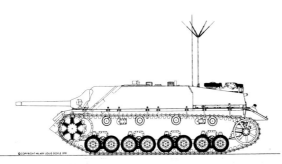

Jagdpanzer IV - May-June 1944 Page 124

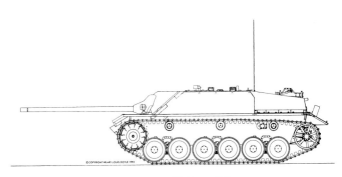

Panzer IV long (E) Page 170

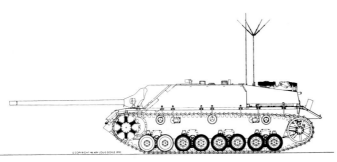

Panzer IV/70 (V) - Sept. 1944 Page 141

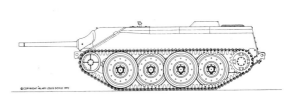

Jagdpanzer E 10 Page 189

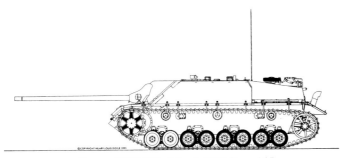

Panzer IV.70 (V) - March 1945 Page 145

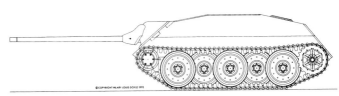

Jagdpanzer E 25 Page 192

Foreword

I no longer feel so able to be burdened in terms of health that I alone could bear the responsibility for the quality of historically necessary research that our readers expect. But I was in a position to prepare my longtime partners **Hilary L. Doyle** and **Thomas L. Jentz** for this responsibility at the right time. The first result, which lies before you now, will convince you. This text is the result of long years of hard work by Thomas L. Jentz, who has remarkable talent for locating relevant documents. His ability to make historical connections logically also promises surprising and new insights into the individual areas of German military motorization in the future.

In addition, his knowledge also allows an extension of the contents, which were, for the most part, technically oriented in the past.

I have reserved for myself the possibility of extending the text and the translation.

The recognition of Hilary Doyle's outstanding work has been expressed over and over by our readers. From the measurements of the vehicles to their reproduction in drawings, he has set his own standards. The number of "stolen copies" of his drawings that turn up all over the world speaks its own language.

The American-Irish-German team of authors that has proved to be so successful to this day will also share indisputable facts and information on this fascinating subject in the future.

We would like to thank Messrs. Karlheinz Münch, Ivo Petr, Alexandr Stefan, and Hans Ström, who have provided us with hitherto unpublished drawings and pictures.

Walter J. Spielberger

Introduction

As a fighting instrument of the Artillery, the assault gun system of weapons wrote its own history. This volume now presents—as a counterpoint to the assault gun—the light **Tank Destroyer** in the light of documentation. The antitank units of the German *Reichswehr* and *Wehrmacht* were motorized as a matter of principle. They introduced the entire range of vehicles—from m.gl. Pkw (Kfz. 12) through makeshift versions of light stock trucks to the series-produced limber (Kfz. 69).

Along with these wheeled vehicles, they used the especially suitable halftrack towing tractors. All of these vehicles, though, required too much time to take up positions, quite aside from their vulnerability to enemy fire for lack of armor.

These problems were temporarily solved by the speedy introduction of self-propelled tank destroyer mounts on the chassis of mustered-out tanks.

They remained a satisfactory temporary solution—their general makeup and armor protection left many wishes unfulfilled—until the introduction of the tank destroyer.

Only they provided the desired balance between firepower, mobility and protection. This volume places their weapon-technical

In 1935 the Chief of the General Army Office ordered the 14th companies of the infantry regiments—the antitank companies—to be motorized.

The towing vehicles used for the 3.7 cm antitank gun at first were mainly medium offroad-capable personnel cars with towing apparatus (Kfz. 12). This one is a Wanderer W 11.

At the beginning of World War II, the *Wehrmacht* made towing vehicles of confiscated light civilian trucks (like this 1-ton Opel). They replaced the motorized limber with towing apparatus (Kfz. 69), or the medium offroad-capable personnel car (Kfz. 12) for towing the 3.7 cm antitank gun.

The standard towing vehicle for the 3.7 cm antitank gun was the limber vehicle (Kfz. 69), Krupp types L 2 H 43 and L 2 H 143.

The six basic types of halftrack towing vehicles (1, 3, 5, 8, 12, and 18 tons) proved themselves as towing vehicles and self-propelled mounts for all antitank guns. This is a Krauss-Maffei KM m 11.

The self-propelled tank destroyer mounts on the chassis of no longer current tanks formed, until the war's end, a usable, though only makeshift solution as antitank vehicles.

Comparison of a self-propelled antitank vehicle and the final form of the tank destroyer (here *Jagdpanzer* 38).

aspects in the foreground for observation. While in the first war years the action of the 3.7 cm and 5 cm antitank guns very quickly showed the limits of this weapon's performance, the deep gap that appeared at the coming of the Russian T-34 tank compelled changes. Thus, the step was made to the best antitank weapons of World War II. The 7.5 cm Antitank Cannon L/70, and above all the 8.8 cm L/71, gave the German fighting forces absolute superiority over all Allied tanks. These new weapons also were a perfect fit for the concept of the tank destroyer. The basis for this development remained, with few exceptions, a series-produced armored vehicle chassis. By omitting the tanks' usual turning turret, the result, with the same fighting weight, was the possibility of improving armor protection and housing a larger caliber weapon.

The necessarily defensive character of German warfare in the last years of World War II brought about the building of more well-armored, high-firepower defensive vehicles as offensive battle tanks. Industry reacted quickly, and found useful solutions that were realized, in present-day terms, in remarkably short times. The combat vehicles that grew from this standpoint were outstanding representatives of their type.

All vehicles of this type were called "*Sturmgeschütz*" (assault gun) when the Artillery named them, and "*Panzerjäger*" (tank destroyer) when the armored troops baptized them. The name for the title of this book was chosen to refer to both the older type of lightly armored self-propelled antitank gun mounts, which were also called "*Panzerjäger*," and the newer type of heavily armored tank destroyers, which were lastly called "*Jagdpanzer*." The titles used for the individual chapters are in every case the last official names of these tank destroyers.

The development of the designations for each individual type can be found in the chapters. During the developmental time, various descriptions were used at the Führer's conferences, official discussions, and on orders from *Waffen-Prüfwesen* 6. The right to make the official designation of armored vehicles belonged to the "In 6" department of the Inspector of the Fast Troops, which was subordinate to the General Army Office. The often-changed names and designations of each tank-destroyer type in the last war years were results of disagreements between Guderian and other generals of the OKH.

This volume covers the hitherto not described knowledge of light armored vehicles gained in the course of recent years. Originally the story of the German tank destroyer was to be handled in a single volume. The dramatic political changes of the most recent years, though, opened formerly inaccessible sources, from which, above all, hitherto unknown photographic material could be gained. Therefore, we soon decided to divide the work into two volumes. The heavy tank destroyers will follow in another volume (*Heavy Jagdpanzer*), which also includes the shared bibliography and all the addenda.

In this volume the reader will find the fascinating broad spectrum of modern light combat vehicles portrayed in detail and comprehensively.

The technical background is enhanced by a richness of actual combat reports, which leave nothing to be desired in terms of clarity.

Jagdpanzer 38

Development

The production of assault guns by the Alkett firm was decisively interrupted when Allied bombers dropped 1424 tons of explosive and incendiary bombs on Berlin on November 26, 1943. Because of this damage, the OKH investigated the possibility of developing assault-gun production at the Böhmische-Mährische (Bohemian-Moravian) Machinenfabrik (BMM),* in Prague. Hitler agreed with the suggestion to devote the production capacity of the BMM firm completely to the building of a new light tank destroyer. The suggested 13-ton vehicle was to have an extraordinarily high top speed of 55 to 60 km/h, so as to equalize to some extent the thinner frontal armor. The side armor plates were only thick enough to offer protection against grenade splinters.

* Formerly Ceskomoravska Kolben Daneck (CDK).

The picture shows the compact form of the hull of *Panzerkampfwagen* 38 (t).

On December 17, 1943, the developmental drawings for the 13-ton assault gun were presented. It was to consist of components of the old *Panzerkampfwagen* 38 (t) and the newer version of a design for a reconnaissance vehicle. Hitler declared emphatically that he regarded this as the best solution for the use of the BMM factory's capacity. Plans were developed in which a recoilless primary weapon was to be introduced in the final version of this vehicle. But since the development and testing of the recoilless weapons were not yet finished, parallel plans were developed,

foreseeing the use of the 7.5 cm *Panzerjäger* Cannon, which was already in series production.

The developmental work took place under heavy time pressure; on January 24, 1944, a full sized wooden model was finished, and was presented to the Army Weapons Office two days later. The office decided to use this same 7.5 cm tank-destroyer cannon, which was also planned for use in the *Jagdpanzer* IV. Colonel Thomale quickly ordered three *Jagdpanzer* 38, which were to be available for troop testing as of March 1944.

Panzerkampfwagen 38 (t)

Development of Designations for the *Jagdpanzer* 38

Leichter Panzerjäger auf 38(t)
WaPrüf 6   1/7 and 2/28/1944
Pz.*Jäger* 38(t)
KTB, GenStdH/Gen.d.Art.   1/18-4/16/1944
Instead "*Sturmgeschütz neuer Art*", le. Pz.Jäger (38t)
Gen.Insp.d.Pz.Tr. to OKH/WaPrüf   1/28/1944
Leichtes Sturmgeschütz auf 38(t)
Führer's conference   1/28/1944
Panzerjäger 38 **für 7.5 cm Pak 39 (L/48)(Sd.Kfz 138/2)**
K.St.N. 1149   2/1/1944
Le.Pz.Jg.38t
Gen. Insp.d.Pz.Tr. Files   3/4/1944 to Oct. 1944
7.5 cm *Panzerjäger* 38(t)
Chef.H.Rüst.u.BdE, Wa.Abn.   4/6-7/31/1944
Stu.Gesch.38(t)
Chef.H.Rüst.u.BdE, Wa.Abn.   4/6-6/6/1944
Stu.Gesch.n.A. mit 7.5 cm Pak 39 L/48 auf Fgst.Pz.Kpfw.38(t)
"Overview of the Armament State of the Army",
Chef.H.Rüst.u.BdE/Stab Rüst III.   5/15-10/15/1944
Le.Pz.Jäg.38(t)
GenSTdH/General der Artillerie War Diary   6/7 & 7/30/1944
Stu.Gesch.38(t)
GenSTdH/Org.Abt. Report   6/12 & 6/28/1944
l.Pz.Jg.38(t) WaPrüf 6   6/23/1944
Le.Pz.Jg.38(t) mit 7.5 cm Pak L/48 auf Fgst Pz 38t
as
"le.*Panzerjäger* 38t"
Gen STdH/Org.Abt./Gen.Insp.d.Pz.Tr.   9/8/1944
Called by the troops: ***Jagdpanzer* 38**

Called by regulations:
***Jagdpanzer* 38 Ausf.**
GenSTdH/Org.Abt./Gen.Insp.d.Pz.Tr.   9/11/1944
Pz.*Jäger* 38(t)
later name probably ***Jagdpanzer*)**
GenSTdH/General der Artillerie War Diary   9/12/1944
***Jagdpanzer* 38**
Gen.Insp.d.Pz.Tr. Files   10/19/1944-4/6/1945
***Jagdpanzer* 38 D652/63**   11/1/1944
***Jagdpanzer* 38 und *Panzerjäger* 38**
(7/5 cm Pak 39 (L/48)) (Sd.Kfz. 138/2)
K.St.N. 1149   11/1/1944
***Jagdpanzer* 38, *Panzerjäger* 38" Overview of the Armament
State of the Army"**
(m. 7.5 cm Pak 39 L/48)(Sd.Kfz.138/2)
Chef H Rüst.u.BdE/Stab Rüst III.   11/15/1944-3/15/1945
***Jagdpanzer* 38** Wa A/Wa Prüf 6   11/17 & 12/19/1944
Explanation of the name **"Hetzer"**
The expression comes from the troops and indicates the ***Jagdpanzer*
38**.
Gen.Insp.d.Pz.Tr. (12/4/1944)

Specific Features

The first wooden models already showed the final low body of the
Jagdpanzer 38. All the armor plate was angled sufficiently in the
direction of oncoming shots. The upper front armor plate was 60
mm thick, with an inclination of 60 degrees to the vertical, while
the thickness of the lower front armor plate was 60 mm at 40
degrees.

Wooden model of *Jagdpanzer* 38, full size, with short fighting compartment. Angled forward side armor. Conical "pig's head" shield, 7.5 cm cannon with muzzle brake.

The 20 mm upper and lower side armor was angled at 40 to 50 degrees. The 7.5 cm Pak 39 was attached to the upper armor inside, a basic difference from the older type of assault gun. While the primary armament there was mounted in a fixed lower mount and a limited-turning upper mount, the new type *Jagdpanzer* had a cardan frame holding the gun built into a carrier rack on the front wall of the armor plate. Without this hanging mount, the weapon could not have been used for lack of space. The installation of the gun far right of the center line gave a very limited traverse field of only 5 degrees to the left and 10 degrees to the right—instead of the 15 degrees to either side foreseen by the original design. The unusual gun attachment also meant that the running gear on the right side had to carry 850 kilograms more weight than the left side. Additional armament in the form of an all-round mount for a machine gun on the roof was foreseen.

Another 1:1 scale wooden model, with complete bow plate, side aprons, new-type weapon mantlet, and no muzzle brake.

A changed 1:1 scale wooden model with a lengthened fighting compartment...

...machine-gun mount on the roof, and cast steel frame for the driver's window. The gratings for the cool-air intake were not kept in series production.

Jagdpanzer 38 Prototype
March 1944, chassis no. 321001 (drive wheels lightened by piercing—side aprons not yet attached—ball mantlet attached by bolts—opening in the upper bow plate—cast towing hooks—protecting plate angled downward).

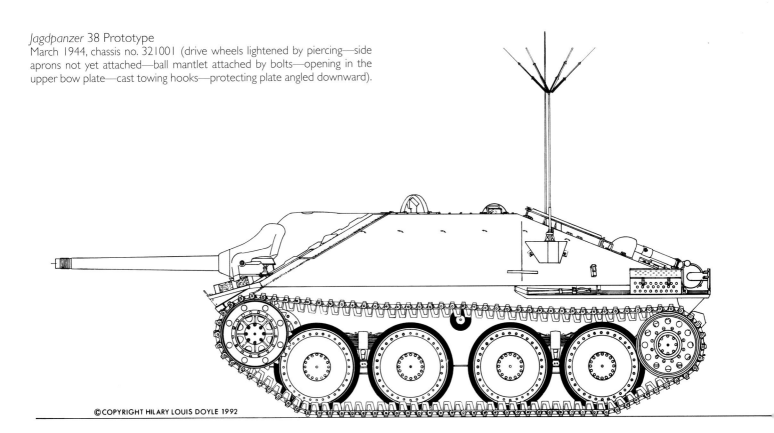

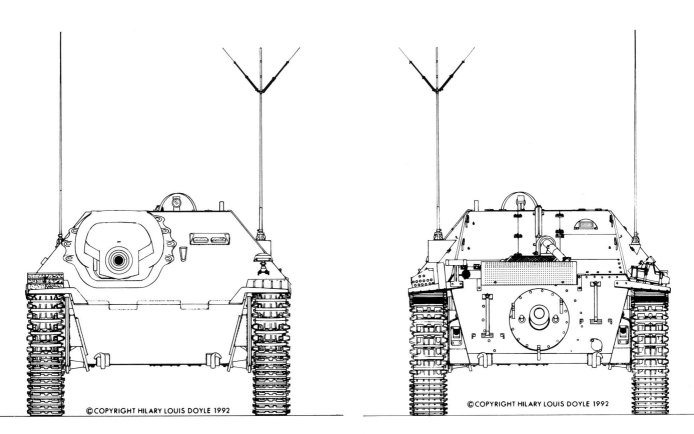

The viewing possibilities were very limited. Two periscopes for the driver and an "Sfl.ZF 1a" for the primary armament for the aiming gunner were planned. The periscope sight for the machine gun and the fixed periscope in 9:00 direction for the loading gunner were supplemented by an Sf 14 Z shear telescope for the commander. When all hatches were closed, the crew on the right side of the *Jagdpanzer* 38 could not see outside.

The powerplant was a six-cylinder gasoline engine (Praga "epa") displacing 7754 cc and producing 160 HP at 2800 rpm.

A semi-automatic five-speed transmission carried the power flow via a Wilson clutch linkage on the front transmission. The

running gear consisted of normal drive wheels and leading wheels of the 38(t) tank series, but had four road wheels with a greater diameter and a single jack roller on each side. These wheels and roller came from the new-type Panzer 38(t) program. The top speed was about 40 km/h, and the *Jagdpanzer* 38 was thus well below the calculated design figures.

The production vehicles weighed 16 tons, instead of the originally foreseen 13 tons. Thus, the power train, clutch, and leaf springs were under much greater pressure than had originally been expected. The vehicle proved to be nose heavy, and was thus ten centimeters lower in front than in back.

Facing pages: Steel replaces wood. Test frame made of steel.

Panzerkampfwagen 38 (t) new type.

On June 25, 1944, the following improvements were suggested to solve these problems:

1. The armor was changed to attain a balanced overall weight.
2. New drive devices were developed.
3. To improve driving, thicker leaf springs were installed to stabilize the unevenly burdened suspension.

Jagdpanzer 38 (Chassis no. 321002). The gun's inner mantlet is not yet installed, and a jack roller is still missing.

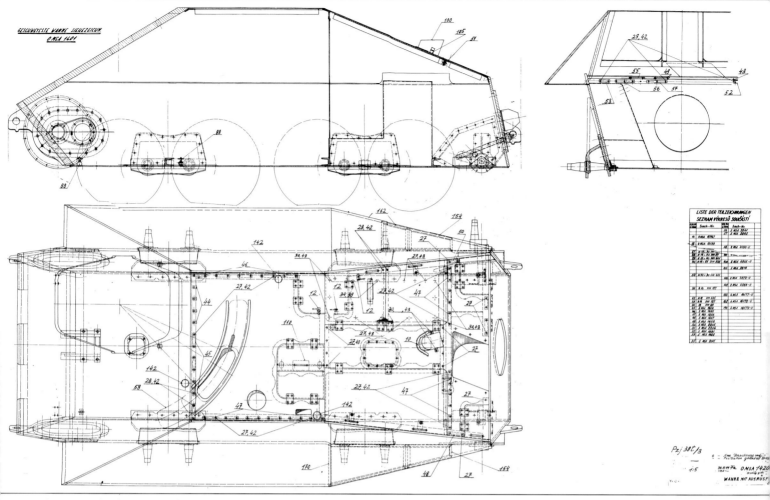

Panzerjäger 38t/3. Drawing of the hull, design state of 11/30/1944.

Although the development took place under heavy time pressure, and the entire design was carried out in the record time of less than four months from the first design to the final prototype, it is striking how few problems turned up. As with all other vehicles, improvements were constantly being made during production, in order to improve the vehicle's performance. Among the changes easily seen from the outside are the following:

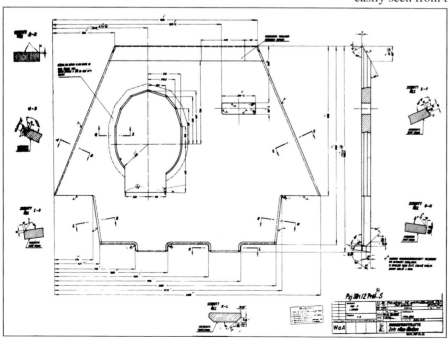

The upper front armor plate.

April 1944:

The towing hooks were removed. They were replaced by a lengthening of the side hull panels, in which openings were bored to create towing eyes.

The weight of the gun mantlet was cut by making the flange smaller.

To save production time, the openings that previously had been bored in the outer rings of the drive wheels were omitted.

The armor protection for the all-round machine gun was shortened, in order to avoid any touching of the head of the Sfl.ZF by the armor.

April 1944: One of the first *Jagdpanzer* 38s in front of the Bohemian-Moravian Machine Works factory in Prague. This is the command-tank version.

The roof machine gun was not yet delivered.

One of the first production vehicles is seen during testing.

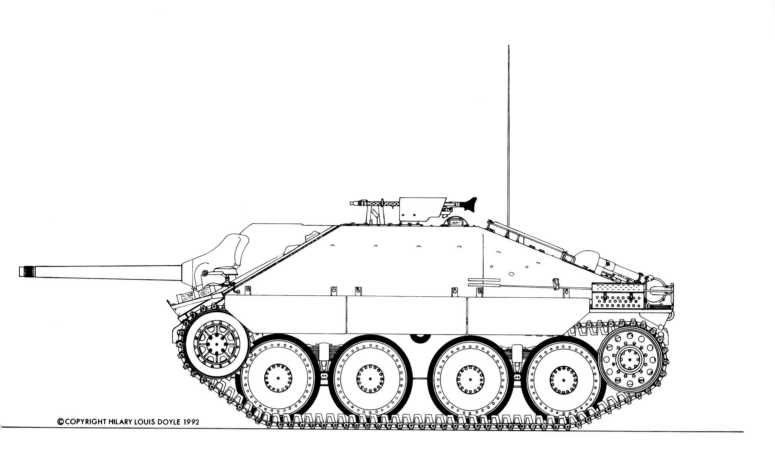

© COPYRIGHT HILARY LOUIS DOYLE 1992

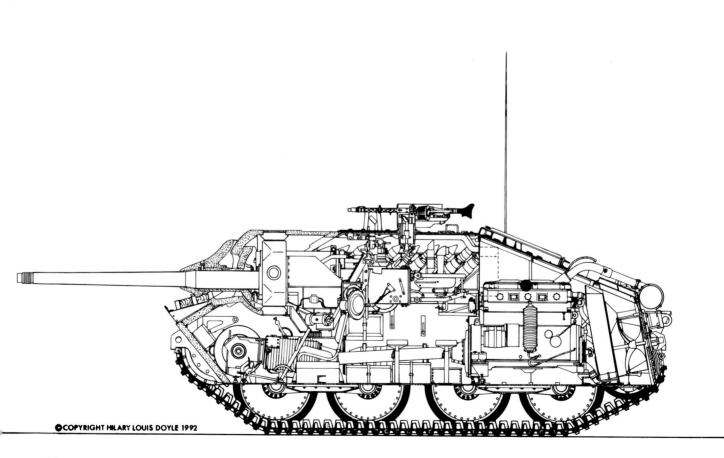

© COPYRIGHT HILARY LOUIS DOYLE 1992

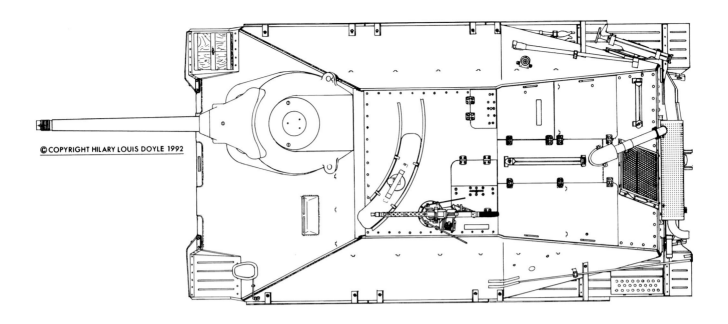

Jagdpanzer 38
May 1944 - Chassis no. 321042 (no holes in the drive wheels - Ball Mantlet III - heavy "pig's head" shield - remote-control machine gun on the roof - cast exhaust pipe joint - side aprons - towing eyes - recoil protection can be folded away - Sfl ZF Ia visor - handwheel for 30-degree traversing machine - towing apparatus across the entire vehicle - one-piece protecting plate).

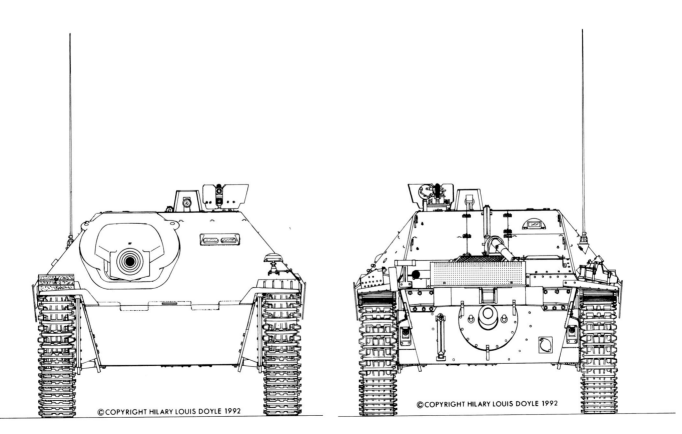

May to June 1944

To improve the means of entering without having to open the large hatches on the rear top plate, three smaller openings were introduced, as follows:

a) a hatch for the commander, opening to the rear;
b) an opening at the lower right for filling with coolant;
c) an opening at the lower left for filling with fuel.

The heat shield around the muffler was eliminated.

Three short cylinders with threading, called "Pilze" (mushrooms), were welded to the roof for attaching the two-ton makeshift crane. This crane was used in order to lift heavy parts, such as the gun, motor, or other power aggregates, during servicing or repairs.

Jagdpanzer 38, end of June 1944: The vehicle's nose heaviness is obvious. The ground clearance was up to 100 mm less in front than in back...

...and the heavy towing apparatus was still used at that time.

August 1944:

A lighter outer and inner gun shield were introduced. Thus the weight decreased about 200 kilograms.

Road wheels with larger-diameter discs and narrower rims were introduced. At first the new disc wheels had 32 holes bored for attaching screws, but in fact only 16 screws were used.

To shorten the work of assembly, as of August 1944 a whole series of changes were made to the pulleys. The following changes were made, in the order of their introduction:

- Reduction to six openings on the original flat upper surface;
- Welded spokes with eight openings on a flat surface;
- Eight openings on a flat surface;
- Pressed ribs on a curved surface with six openings;
- Six openings on a curved surface;
- Four openings on a curved surface.

Jagdpanzer 38 (chassis no. 321003) during factory testing.

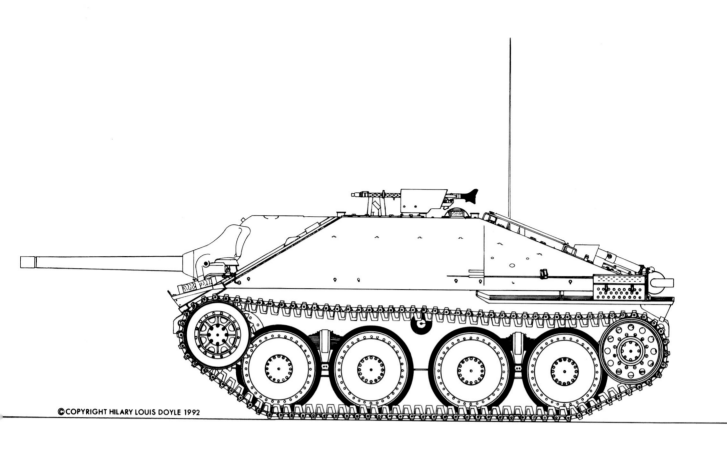

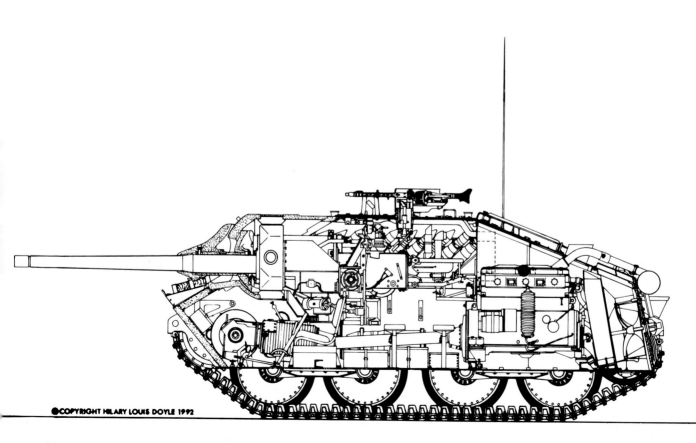

Jagdpanzer 38
July 1944 - (Ball Mantlet IV- "Mushrooms" for 2-ton crane - changed *ersatz* antenna mount-welded sheet-metal exhaust pipe joint - inner and outer shield armor weight reduced - smaller commander's hatch - experimental flap for fuel filler cap - no exhaust heat shields - curved rail for visor shortened).

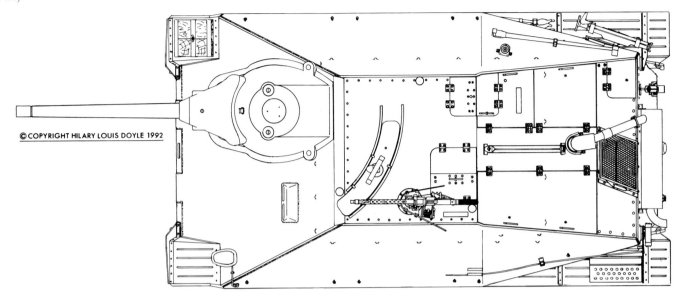

© COPYRIGHT HILARY LOUIS DOYLE 1992

The machine-pistol opening to the driver's right was closed, since the machine gun was now installed on the roof. Behind the shear telescope for the commander, a periscope was installed.

September 1944:

The ends of the side aprons were curved inward. They tore off easily when the vehicle touched objects.

To decrease the number of broken leaf springs because of the overburdened suspension, the front set of springs was made of 16 leaves, each 9 mm thick. The rear 16 leaf springs remained 7 mm thick.

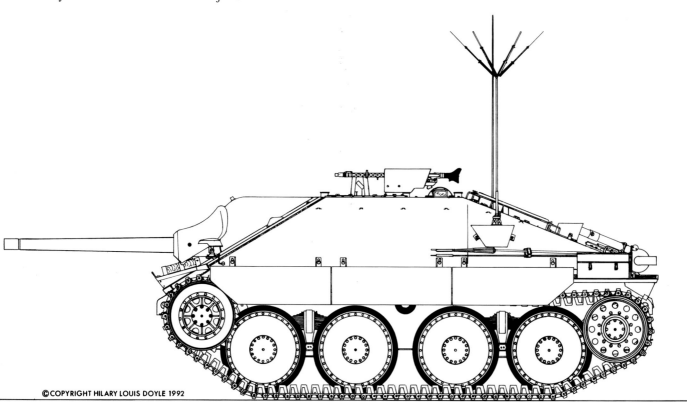

©COPYRIGHT HILARY LOUIS DOYLE 1992

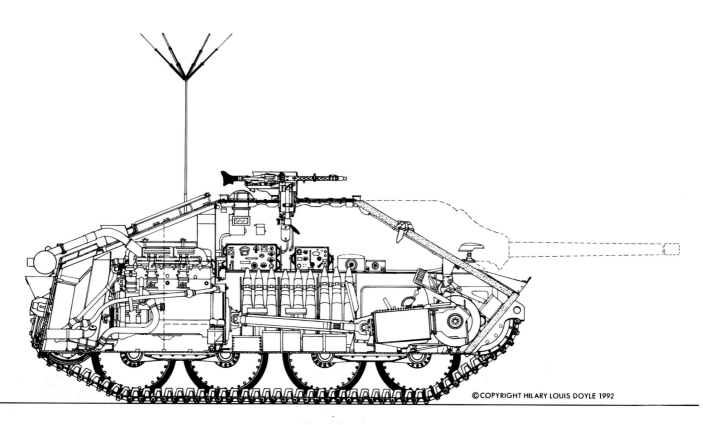

©COPYRIGHT HILARY LOUIS DOYLE 1992

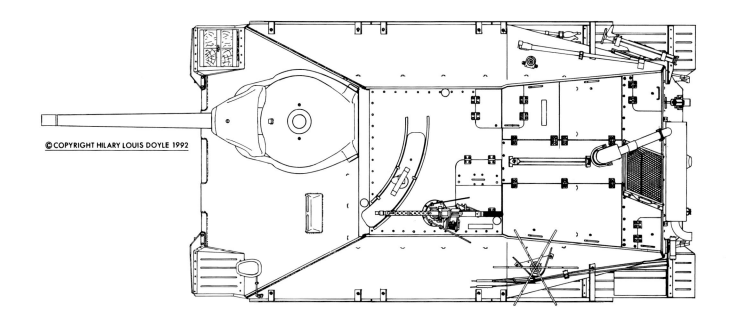

Jagdpanzer 38 Command Tank
September 1944 - (Star antenna D - ball mantlet V - improved "pig's head" shield - new road wheels with larger rims (32 bolts) - new springs - Radio Fu 8 and Fu 5 placed in a niche on the left side of the vehicle - flaps for filler caps on both sides of the engine - compartment cover).

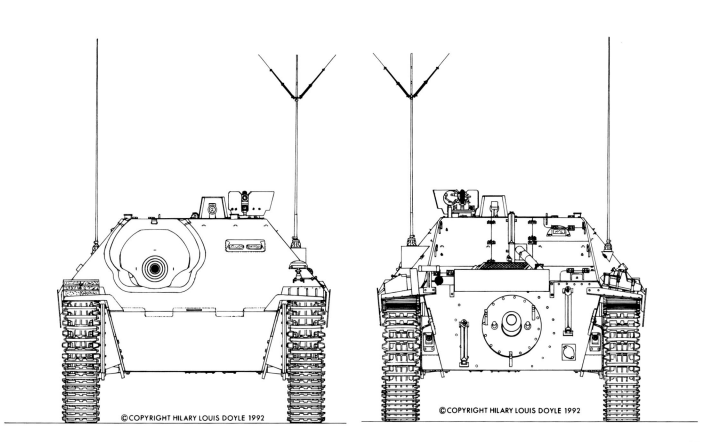

October 1944:

Tank shells penetrated the driver's periscope bracket. This happened mainly when they hit deep on the bow plate, then slid up the armor plate and were caught on the projecting housing. The armor protection was therefore removed, and the periscope housed in openings even with the front armor. A sheetmetal shield prevented rain or sun from affecting the driver's view.

The screws that were to hold the rims to the discs of the road wheels constantly came loose. New road wheels, on which rivets replaced the screws, were introduced.

In order to avoid ignition failures and glowing exhaust pipes, which betrayed the vehicle's presence at night, a flame-extinguishing exhaust system was used in place of the original type.

As of 1945, rings were welded to the upper front of the hull and the sides of the body, so camouflage material could be attached to the vehicles.

The lengthened side plates, which had holes bored in them for use as towing eyes, were either reinforced by added plates or left off, replaced by U-shaped towing loops at the front and back of the vehicle.

In addition, the following **internal** improvements were made:

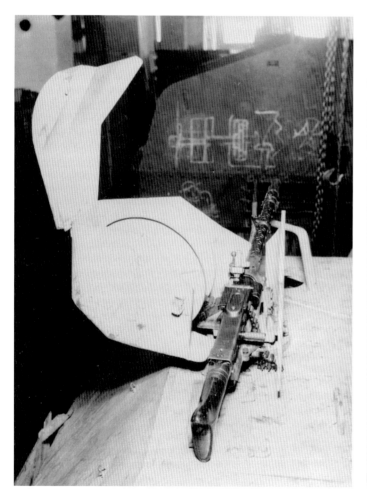

An experimental drum magazine for the MG 34 was attached to the roof.

One of the first twenty finished *Jagdpanzer* 38 (April 1944). The side armor plates were lengthened and shaped into towing eyes in front. A Type II Ball Mantlet is built in. The mantlet was held only by two bolts. On the sides, the mantlet was better protected by two welded-on panels.

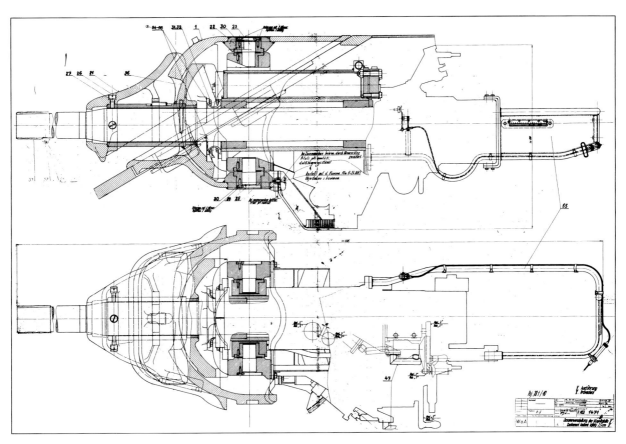

Composition of the ball mantlet, Type V, is shown in a drawing from July 1944.

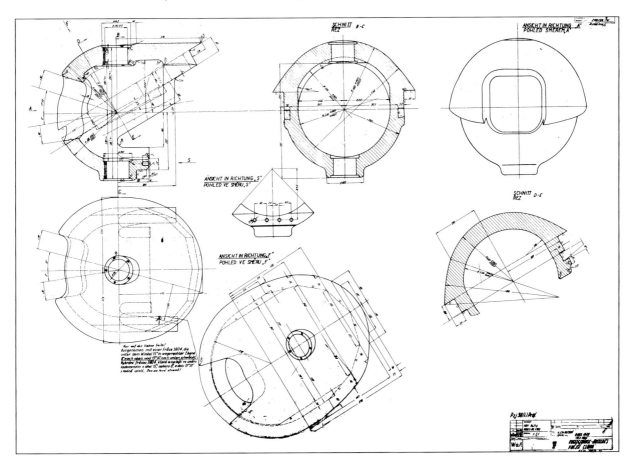

Details of the ball housing are shown in this drawing from June 8, 1944.

August 1944:

Two handholds were welded to the inside of the vehicle over the driver's seat, to allow him to get out more quickly.

October 1944:

A spring equalizer was installed on the breech of the gun to simplify the elevation of the cannon. This became necessary after roller bearings were used for the shield trunnions of the primary armament for lack of roller bearings.

A larger filler cap with a deflector panel shortened the fueling time.

A Solex hand pump replaced the unreliable electric fuel pump.

Head protection was installed inside on the commander's hatch cover.

Rear view of one of the first twenty *Jagdpanzer* 38. The commander's hatch was made smaller experimentally.

November 1944:

To be able to carry more shells, the box that held means of vision to the right of the commander was moved. In this way, room was made for five more 7.5 cm Pak shells.

An improved water pump was installed.

An improved divider channel was installed in the firewall, improving the heating for the crew.

A heating plate was installed for the battery, to prevent freezing.

One of the first twenty *Jagdpanzer* 38, assigned for training in April 1944, was destroyed during the combat at Prague in May 1945. The vehicle had winter tracks on.

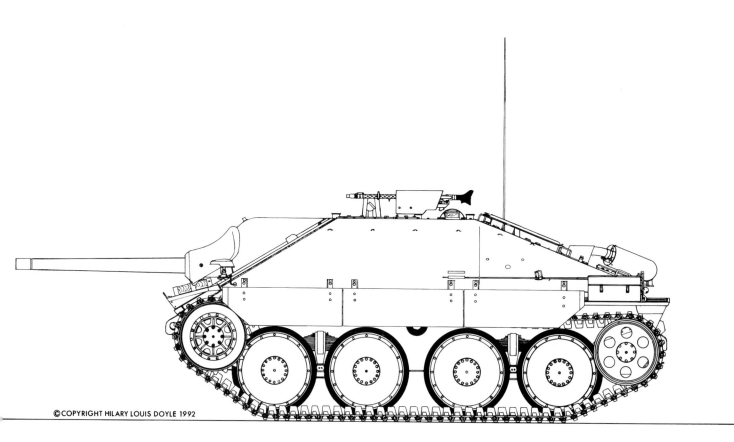

Jagdpanzer 38
December 1944 - Chassis no. 3211 (new driver's scope - new 16-bolt road wheels - new leading wheel with six openings - side aprons bent inward at front and back - flame extinguisher on the exhaust - new track links - Ball

Mantlet V - Sfl ZF 1c scope - recoil protection no longer folds away - new gun mantlet - changed towing eyes, the rear ones more deeply mounted - improved lubricating system for the motor - new periscope - protective panel simplified again - new track-tightening device - fuel tanks and tools in the engine compartment).

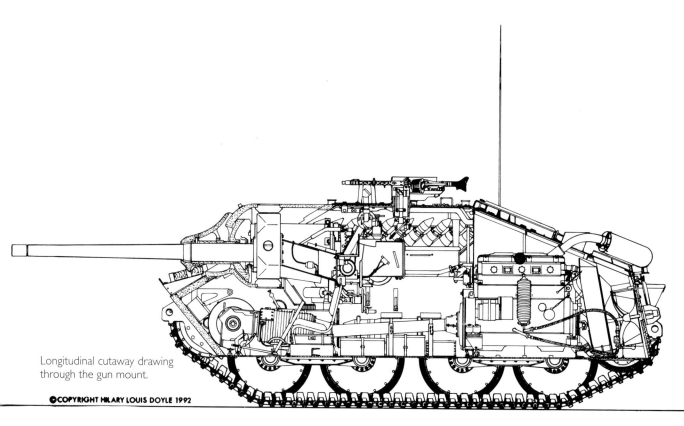

Longitudinal cutaway drawing through the gun mount.

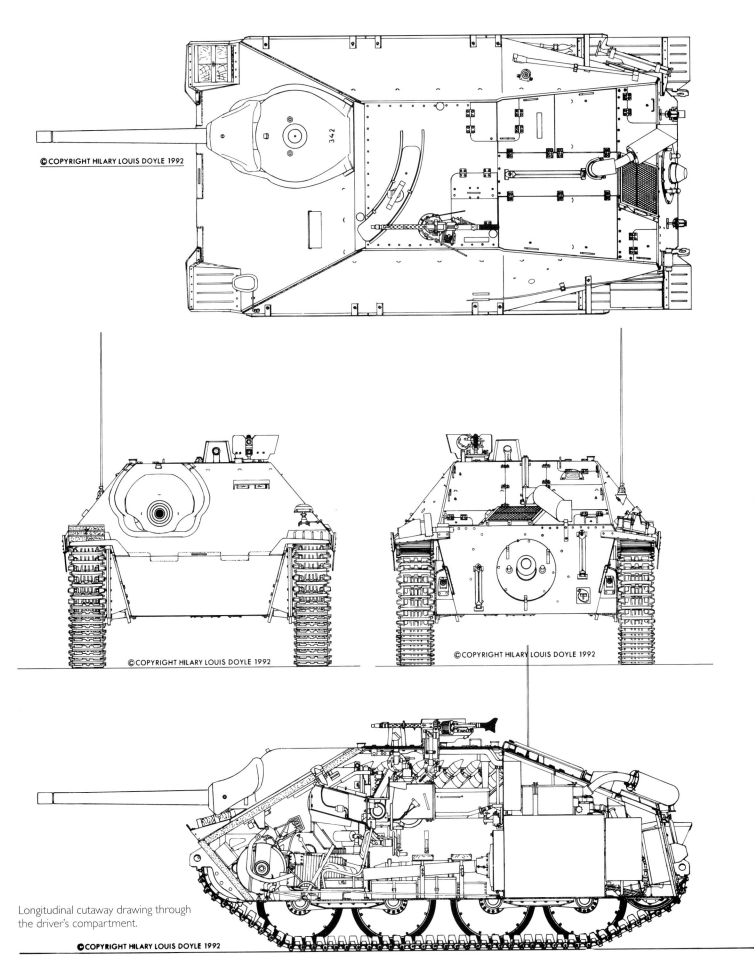

© COPYRIGHT HILARY LOUIS DOYLE 1992

© COPYRIGHT HILARY LOUIS DOYLE 1992

© COPYRIGHT HILARY LOUIS DOYLE 1992

Longitudinal cutaway drawing through
the driver's compartment.

© COPYRIGHT HILARY LOUIS DOYLE 1992

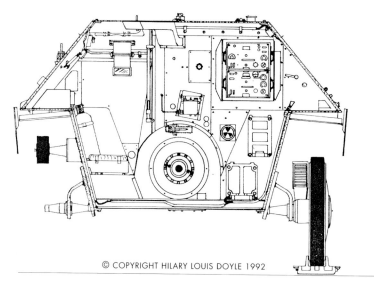

Transverse view, looking toward the firewall between the fighting and engine compartments, showing the Fu 5 radio set.

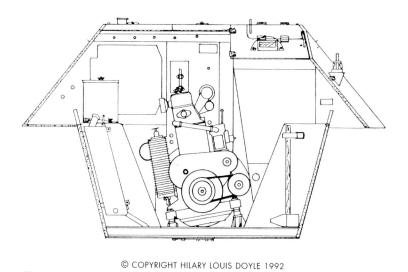

View into the engine compartment from behind. Fuel tanks on both sides, battery box above the left fuel tank.

Other training vehicles that were destroyed in combat in Prague. The rear towing eyes are screwed in. The commander's hatch was made smaller later. The upper vehicle had wider winter tracks.

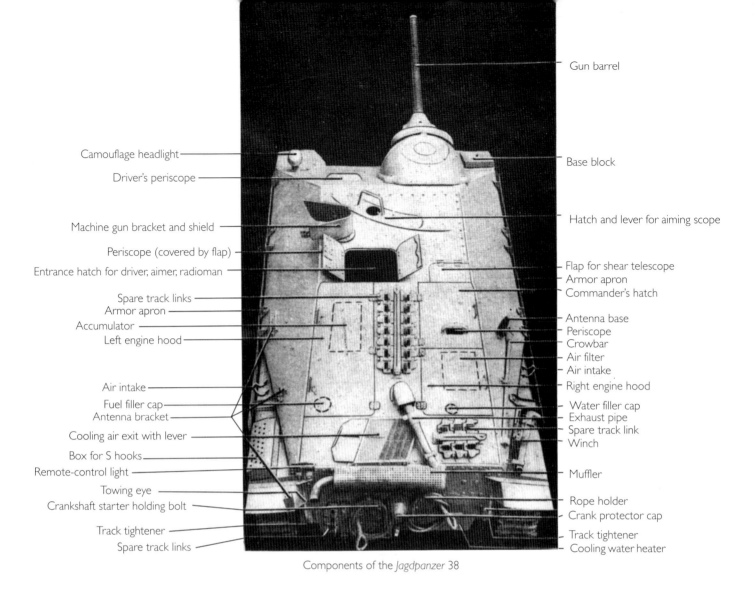

Camouflage headlight
Driver's periscope

Machine gun bracket and shield
Periscope (covered by flap)
Entrance hatch for driver, aimer, radioman
Spare track links
Armor apron
Accumulator
Left engine hood

Air intake
Fuel filler cap
Antenna bracket
Cooling air exit with lever
Box for S hooks
Remote-control light
Towing eye
Crankshaft starter holding bolt
Track tightener
Spare track links

Gun barrel

Base block

Hatch and lever for aiming scope

Flap for shear telescope
Armor apron
Commander's hatch

Antenna base
Periscope
Crowbar
Air filter
Air intake
Right engine hood

Water filler cap
Exhaust pipe
Spare track link
Winch

Muffler

Rope holder
Crank protector cap

Track tightener
Cooling water heater

Components of the *Jagdpanzer* 38

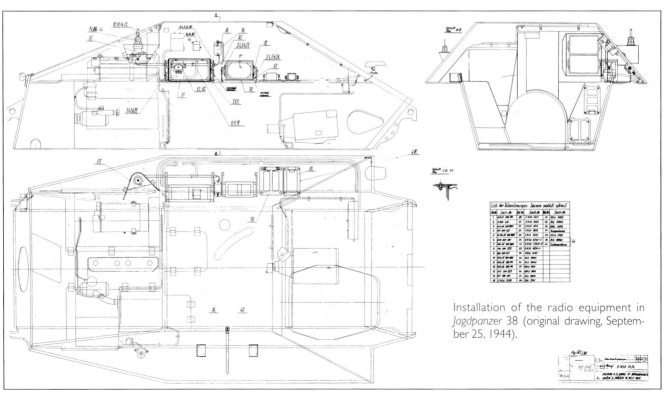

Installation of the radio equipment in *Jagdpanzer* 38 (original drawing, September 25, 1944).

January 1945:

The situation in January 1945 was as follows:

"From the front there now occur serious protests about intermediate gear damage to all types of vehicles. About 200 damages to the 38(t) were reported. For the Panzer IV there were around 500 defective intermediate gears before the 1945 eastern offensive, about 100 for the Panther 370 and the Tiger. General Thomale explained that under these conditions an orderly use of armored vehicles is simply impossible. The troops lose confidence and give up the whole vehicle under the conditions just because of this difficulty. He asks to increase the expenditures on intermediate gears, because calm will set in only in that way."

Because of the limited traversing field of the primary weapon, the vehicle often had to change its position by steering in order to fire on targets. This overtaxed the intermediate gears, which very often ended in a breakdown of the vehicle. In mid-January 1945 a new model was introduced, with a gear ratio of 6.75 and a strengthened 10:80 gearing, which replaced the old Model 6 with 12:88 gearing.

Right: A close look at the ball mantlet, Type III (chassis no. 321042). The armor extensions welded on the sides were later dropped to save weight.

Below: A rear view of the *Jagdpanzer* 38 (chassis no. 321042), on display at Camp Borden, Canada, shows the towing apparatus extending over almost the entire width of the vehicle. The commander's hatch is the series production type.

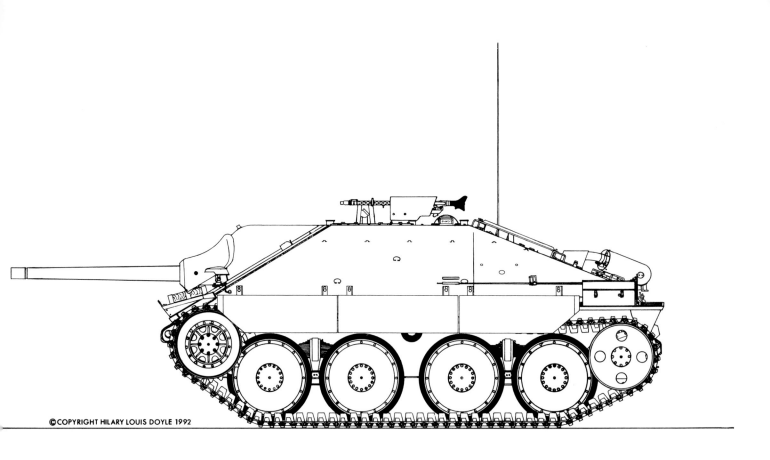

Jagdpanzer 38 - Produced by Skoda.

April 1945 - (Skoda production near the war's end - light leading wheel with four openings - rings on the hull sides to attach camouflage - simplified production - welded-on hinges - changed right engine cover - tools attached to the outer walls - strengthened towing eyes.

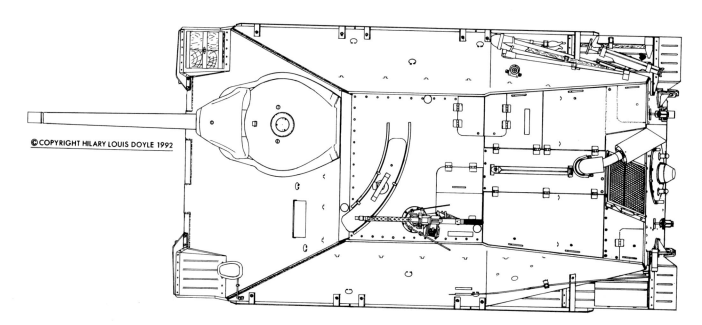

Further Development

Since the required higher performance, compared to the original motor, was possible only by increasing the engine speed, there were problems that influenced the temperature of the engine unfavorably. The WaPrüf 6 requested in May 1944 that the Dr.-Ing. h.c.F. Porsche firm create a new cooling system for the Hetzer chassis. The meager effect, the high fuel consumption, and the size of the former cooling system were criticized. Improvements proved to be insufficient, and a completely new system, based on the experience gained from the Maus engine cooling (Porsche Type 263), was suggested.

But this required major changes to the chassis, including relocating the fuel tanks. The Army Weapons Office did not agree to it. In October 1944 the whole project was given up.

It was foreseen, though, that the development could be brought back to life when the production of the 38 D Diesel version began.

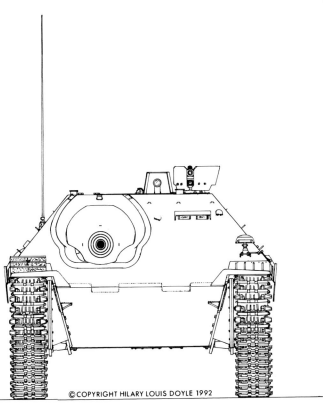

© COPYRIGHT HILARY LOUIS DOYLE 1992

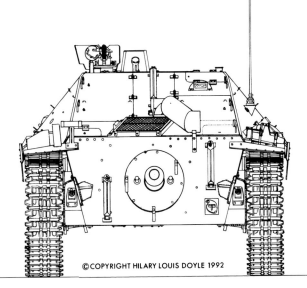

© COPYRIGHT HILARY LOUIS DOYLE 1992

Details of the air filter installation.

Two *Jagdpanzer* 38 made in August 1944, with Type IV ball mantlet and light "pig head" shield.

Porsche suggestion no. 1 for a changed cooling system.

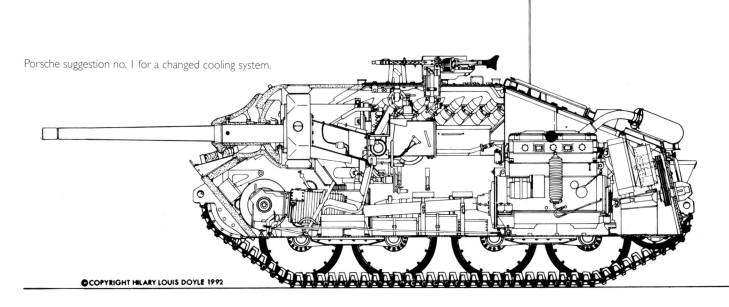

©COPYRIGHT HILARY LOUIS DOYLE 1992

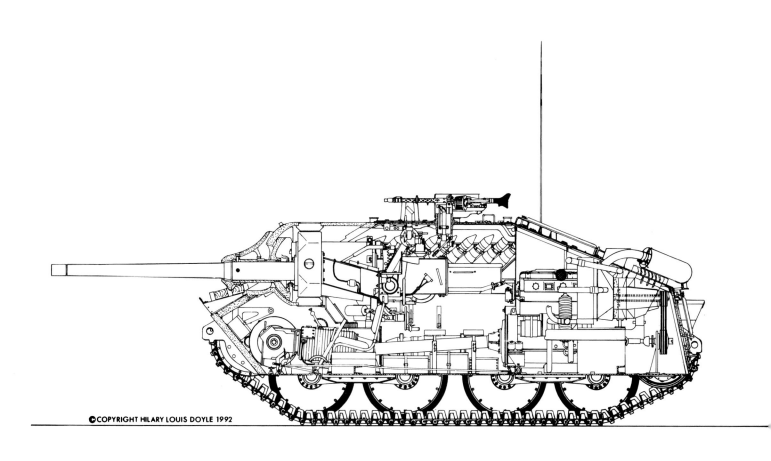

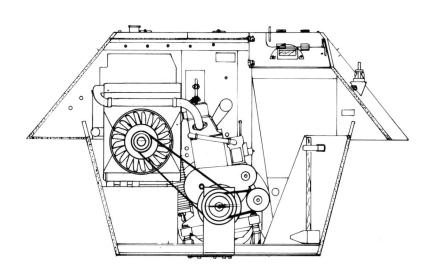

Porsche suggestion no. 2 for a changed cooling system.

Jagdpanzer 38 and Jagdpanzer 38 (D) with Diesel motors.
(Drawings and pictures on pages 67ff)

On 19 March Guderian requested an immediate change to Diesel engines, because of the prevailing shortage of gasoline. This order was not carried out, though, since it would have caused major delays in *Jagdpanzer* 38 production.

In January 1945, Dipl.-Ing. Michaels of the Alkett firm of Berlin-Borsigwalde reported on the state of the *Jagdpanzer* 38 D project. The increase of torque from 42 to 78 m/kg through using the Tatra Diesel engine gave hope for greater mobility of the vehicle. The simple way of equipping the 38(t) with the new motor had been impossible. The BMM design of the 38(t) would not allow rational series production because of the too-great need for hand

Jagdpanzer 38 (chassis no. 321364 - hull 303 - series 12), produced in August 1944, on display in Axvall, Sweden. Old heavy "pig's head" shield, but with new ball mantlet, Type V. Leading wheels are the light version.

fitting. Unfortunately, only the tracks and road wheels of the old vehicle could be used after redesigning. The installation of the air-cooled diesel engine brought a series of new questions about fitting. The former weight of 14.5 tons had risen to 16.7, including 62 rounds of ammunition (1000 kg). The gearbox, by doubling the torque from that of the Wilson gearbox, had required only 15% of excess weight expenditure. The fuel consumption of 76 liters per 100 km on easy terrain gave a range of 510 km with a fuel capacity of 390 liters. Michaels assured the design would allow the best series tank production. In the spectrum of *Jagdpanzer* building, the following variations were foreseen for the 38 D:

- *Jagdpanzer* with 7.5 cm Pak 39 L/48
- *Jagdpanzer* with 7.5 cm PzgK 42 L/70

Until the Tatra Diesel engine was developed, Maybach suggested their HL 64 gasoline engine with fuel injection, producing some 270 HP instead of the 210 HP of the Tatra motor. This development was no longer completed (see drawings on page 73).

Skoda production, ball mantlet, Type IV, camouflage paint by the troops.

Jagdpanzer 38, built at the end of August 1944. Ball mantlet, Type IV, new hull, but heavy "pig's head" shield, camouflage paint by the troops.

Facing pages: *Jagdpanzer* 38 made by BMM, August-September 1944, ball mantlet, Type V, light "pig's head" shield, light-type leading wheels...

...the protective covering of the exhaust system was removed.

Jagdpanzer 38 built in September 1944. New, stronger front springs, side aprons not yet installed.

Interior view from the fighting compartment toward the firewall. Commander's seat at left, at the right rear is the loader's seat, with the aimer's seat in front of it.

Jagdpanzer 38 made by BMM in September 1944, with now standard coolant filler and new road wheels, plus the old hull with high-mounted towing eyes.

Jagdpanzer 38, BMM production, October 1944. New driver's vision setup, new road wheels and new hull, changed paint scheme.

BMM production in October 1944, armored bodies ready for final assembly.

BMM production, October 1944: the picture of this vehicle shows the flame extinguisher mantle on the exhaust pipe. The side aprons are curved inward at the front and back, to prevent being torn off.

BMM production, October 1944, rear view of the vehicle with lighter leading wheels, including reinforced stamped ribs.

BMM production, November 26, 1944 (chassis no. 3211111). Hull from Skoda, Pilsen (no. 1060), 13th series. New leading wheels. On display in Great Britain.

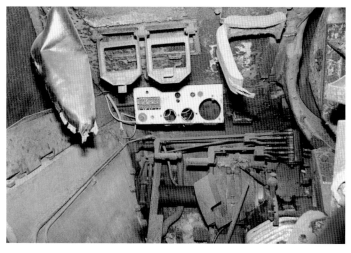

Driver's seat

Ammunition storage in the fighting compartment.

Firewall between fighting and engine compartments.

Room for radios, left side of the vehicle.

View from the aimer's seat.

Engine compartment from left. At left front is the battery box (on the fuel tank).

Skoda production (chassis no. 323211), finished in December 1944. Used as a test vehicle at Kummersdorf since January 11, 1945.

A *Jagdpanzer* 38 made of spare parts and used by the Czechoslovakian forces.

A test vehicle of the Swiss Army. The vehicle, armed with the 7.5 cm Pak 40, was introduced by the Swiss as Type G 13.

Recoilless Armament

The plans for the *Jagdpanzer* 38 with recoilless armament remained at the test stage until the war's end. It did not involve a recoilless weapon on the principle of the "*Panzerschreck*," in which a rocket was fired, but rather the installation of a normal cannon with conventional ammunition, but without recoil apparatus. The recoil power was absorbed by the entire vehicle (tests showed that a part of the recoil power even stressed the heads of the aimer and driver). After the technology still had not been tested and *Jagdpanzer* 38 production must not be interrupted under any circumstances, only 14 *Jagdpanzer* 38 with recoilless guns were built. A standard vehicle was rebuilt by BMM in May 1944. Two more *Jagdpanzer* 38 were turned over to Krupp and Alkett in September for testing. BMM completed a zero series of ten *Jagdpanzer* in December 1944 and January 1945; the last one left the factory in April 1945.

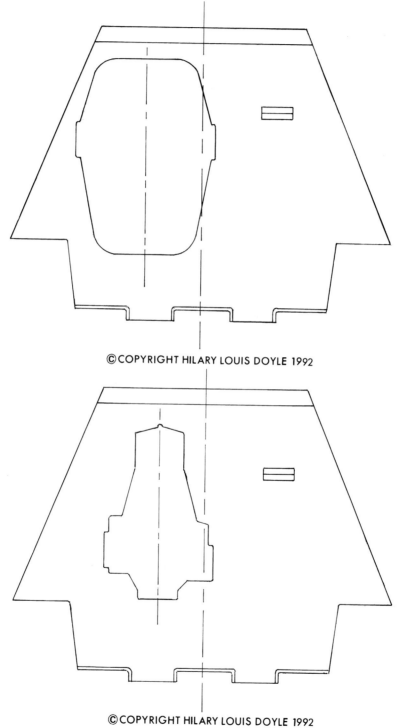

© COPYRIGHT HILARY LOUIS DOYLE 1992

© COPYRIGHT HILARY LOUIS DOYLE 1992

Jagdpanzer 38 - Recoilless weapon
April 1945 - chassis no. 32917
Comparison of the openings for primary weapons with normal recoil (left) and fixed installation. The middle of the gun moves closer to the middle of the vehicle.

Opposite and above: In October 1944 the Bohemian-Moravian Machine Works began to build a 0 series of ten *Jagdpanzer* 38 with recoilless armament. The opening for the gun could thus be made a good deal smaller.

Below: A comparison with the production vehicle (below) shows the different openings for the gun installation. The possibility of placing the longitudinal axis of the gun more in the middle of the vehicle resulted.

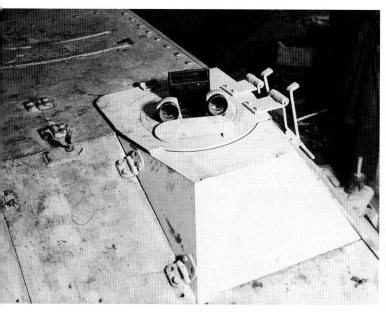

To make more space for the commander, the hatch construction was moved backward. The shear telescope and the periscope were mounted so they could turn. The opening for the shear scope could be closed completely.

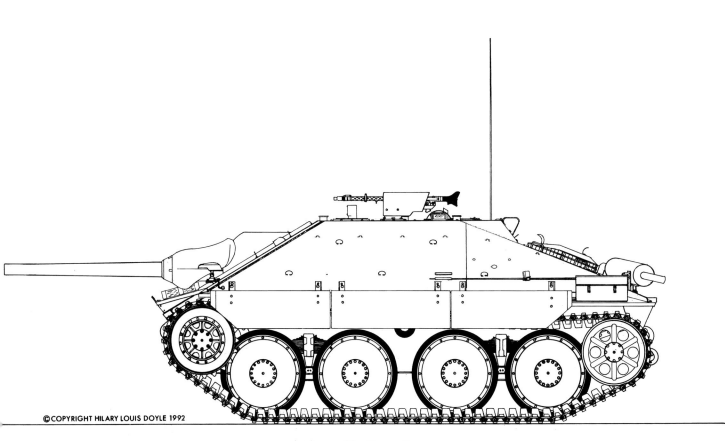

©COPYRIGHT HILARY LOUIS DOYLE 1992

Jagdpanzer 38 with recoilless armament.

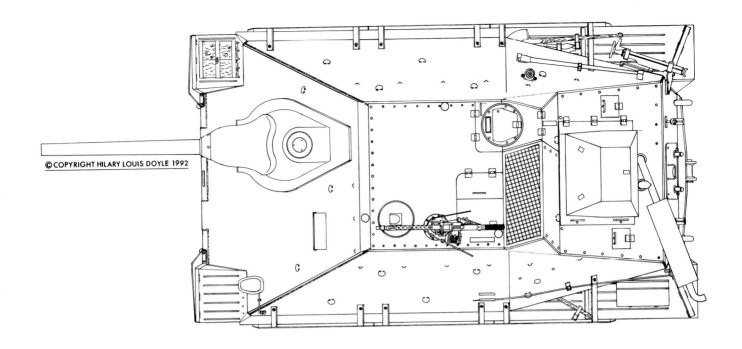

© COPYRIGHT HILARY LOUIS DOYLE 1992

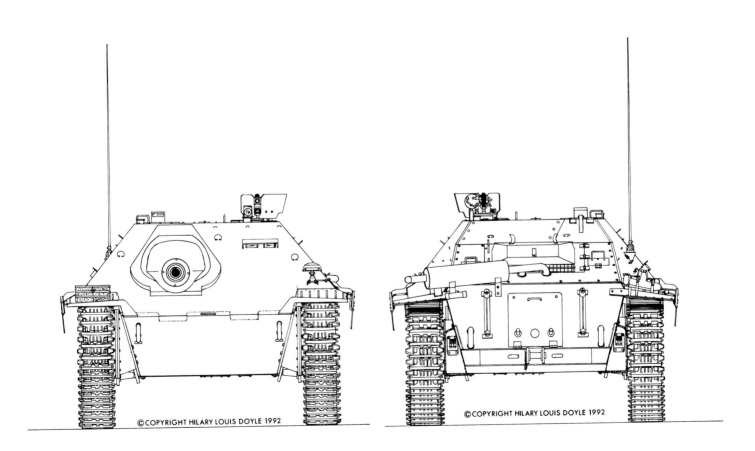

© COPYRIGHT HILARY LOUIS DOYLE 1992

© COPYRIGHT HILARY LOUIS DOYLE 1992

Further changes with fixed installation: WZF 2/2 scope - lengthened headroom for the commander - ribbed leading wheels - turning commander's scope - new engine compartment cover for the Diesel engine - widened track covers and apron attachments - U-shaped towing eyes - completely changed closing panel.

According to an order from the Army Weapons Office on April 21, 1945, Skoda was to produce 500 *Jagdpanzer* 38 with recoilless armament and Tatra V-8 Diesel engine.

Preparations to build the *Jagdpanzer* 38 with the rigidly mounted cannon were also undertaken by Skoda.

On February 19, 1945, the Bergische Stahlindustrie firm of Remscheid answered a Skoda inquiry of December 8, 1944, regarding the delivery of 60 sets for the "fixed" Hetzer as follows:

"60 housings at the weight of 55 kg per piece (price per piece RM 700-), and 60 deflectors weighing 290 kg (RM 340 per piece)."

On March 31, 1945, the Berka Panzer Company was set up as the Berka Testing Department of the Army Weapons Office for the direct defense of this area. It received a *Jagdpanzer* 38 with recoilless cannon, which was stationed in Berka for testing purposes. On the same day, Hitler personally gave the order to destroy this *Jagdpanzer* 38 immediately if necessary, before it could fall into enemy hands.

On April 21, 1945, the OKH Wa Pz Mot, Section IV, notified the Skoda firm:

"You will receive under order number 4911-0210-9004/45 a contract to deliver 500 *Jagdpanzer* 38 with 8-cylinder Tatra Diesel motors* and fixed cannons. This contract runs under the special urgency designation of 'Führer's Need Program.'"

* The air-cooled Tatra V-8 Diesel engine, Type 928, produced 180 HP at 2000 rpm. Overseen by Referat Pz IIIa, tests took place at Berka in June 1944.

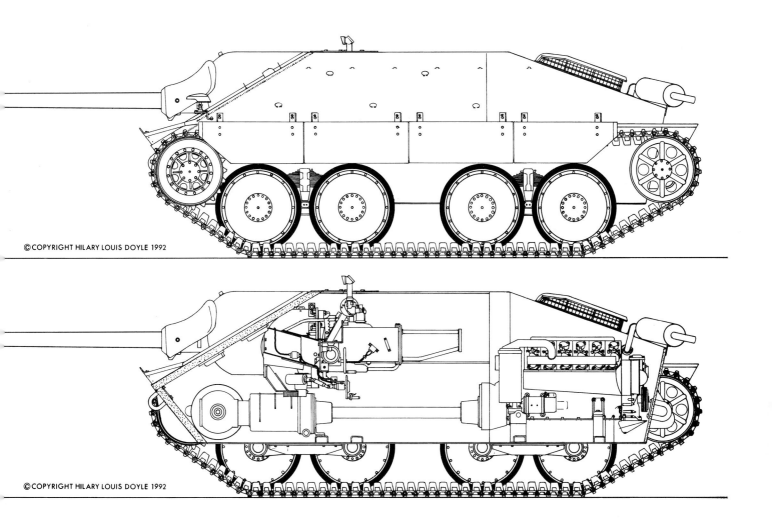

© COPYRIGHT HILARY LOUIS DOYLE 1992

© COPYRIGHT HILARY LOUIS DOYLE 1992

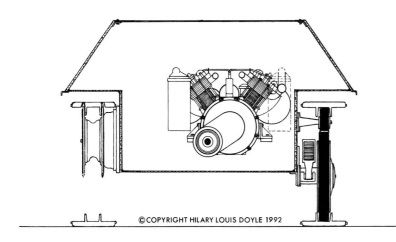

© COPYRIGHT HILARY LOUIS DOYLE 1992

Jagdpanzer 38 D - Prototype
Jagdpanzer 638/29
(Lengthened hull - greater ground clearance - new engine compartment - Tatra Type 103 air-cooled diesel engine - ZF 5-80 transmission - wider hull with vertical sidewalls - central cardan shaft 275 mm over hull floor)

The model vehicle made by BMM had the longer commander's cupola. New towing eyes were used. The side aprons were set farther out from the sidewalls of the hull.

Production

On January 28, 1944, Adolf Hitler stressed the importance of starting and completing production of the "Light Assault Gun on 38(t)" as the most important task of the war industry for the year of 1944. On January 18, 1944, the decision was already made to produce 1000 "Light Tank Destroyer 38(t)" before the ink on the design drawings even dried. An extremely aggressive production was planned; it was supposed to provide for a quick increase to a final goal of 1000 vehicles per month until March 1945.

Jagdpanzer 38 - production at the Bohemian-Moravian Machine Works in June 1944. The ammunition magazines are being installed.

The end of the production line at BMM. The primary armament is installed. At the left front is a command tank. The machine guns to be installed on the roofs are not yet delivered. The openings made for them are temporarily closed by round armor plates (Picture taken 6/19/1944).

Front and rear views of two *Jagdpanzer* 38 made in July 1944 at the BMM factory. The front leaf springs are strengthened. The opening for the commander's position was made smaller. The roof shows welded-on "mushrooms" to hold the two-ton crane. The opening for coolant filling on the right rear engine cover was regarded as an experiment.

1944

Firm	Apr	May	Jun	Jul	Aug	Sep	Oct	Nov	Dec
BMM	20	50	100	200	250	300	350	400	400
Skoda	0	0	0	10	50	100	150	200	300

1945

Firm	Jan	Feb	Mar
BMM	400	450	500
Skoda	400	450	500

Those were overwhelming statistics, when one considers that up to that time no tank factory in the German Reich had built more than 300 tanks per month. The former highest monthly production attained by BMM was 151 units. Skoda, apart from a few prototypes, had not built a single armored full-track vehicle for the *Wehrmacht*.

Facing pages: Production at Skoda, right after the Americans had occupied the factory at Königgrätz. Changed engine compartment covering with welded-on hinges. Strengthened towing eyes. Leading wheels with four openings. The tool was now stowed on the outside of the vehicle.

As ordered, the first three light *Panzerjäger* 38 were finished in March 1944, and accepted by the Army Weapons Office inspector in April. Another twenty followed, and were shown to Hitler on April 20, 1944. After the showing they were sent back to the manufacturer without delay, since they were not yet fully prepared for service; a few parts of their armor were still missing. BMM reached its production goals of 50 tanks in May and 100 in June. This number decreased in July, since the delivery of weapon mounts had apparently been delayed. Although the factory reported that these *Jagdpanzer* were ready for service, and the Army Weapons Office inspectors accepted them, there were enough small problems, such as porous seals in air filters, carburetors, spark plugs, governors, and in the arrangement of the connections between the fuel tanks.

The Weapons Office decreased the production numbers from August to December in order to give the manufacturers more time to deal with these small matters, and deliver vehicles ready for service. Skoda finished the first ten vehicles in July, as anticipated. After that, the firm could scarcely handle the rapidly growing production numbers—especially because of the shortage of experienced workmen.

Four firms received contracts to produce components for the *Jagdpanzer* 38. They were Skoda in Pilsen, BMM, Linke-Hofmann-Busch in Breslau, and Poldihütte in Komotau. With two air attacks on the Skoda works in October 1944, in which 417 tons of explosive bombs were dropped, the firm explained to the Army Weapons Office why the production goals for October could not be met. More than 400 *Jagdpanzer* 38s were produced in November. The Skoda production fell back again in December on account of three air attacks, in which Allied bombers dropped 375 tons of explosive bombs.

The monthly production reached its high point with 434 tanks in January 1945. The industry was not in a position to build the magic number of 500 tanks per month that the Army Weapons Office had required in its need program for 1945. After February 1, 1945, only 2100 additional *Jagdpanzer* 38s were ordered built. The production was to be switched to the *Jagdpanzer* 38 D—a simplified vehicle with a Diesel engine—as of June 1945.

Production fell slightly in February 1945, mainly because of an air attack on Prague. Further production decreases in March and April were caused by the failure of electric power and the first heavy air raid on BMM, in which Allied bombers dropped 378 tons of explosive bombs on March 25, 1945.

Skoda was also hit hard when bombers dropped more than 500 tons of explosive bombs on April 24, 1945.

A few more *Jagdpanzer* 38 were finished in the first few days of May.

It was a noteworthy success for the industry that during the last war years, and under such conditions, they could produce more than 2800 *Jandpanzer* 38.

Two *Jagdpanzer* 38 not finished by Skoda were turned over to BMM on may 31, 1945, by the U.S. Army, finished there, and then shipped to the USA. One of the vehicles (chassis no. 323814) is now at the tank museum in Aberdeen.

Tank components and finished tanks were delivered by railroad.

The Bohemian-Moravian Machine Works experienced their first Allied air raid on March 25, 1945. The picture shows destroyed vehicles on the assembly line.

Bomb damage at the Pilsen works of Skoda. The body production located there was drastically interrupted.

Month	Planned*	WaA Accepted	BMM Built	Skoda Built	BMM Rescue Tanks Built
1944					
March	0	0	3	0	0
April	20	23	20	0	0
May	50	50	50	0	0
June	100	100	100	0	2
July	175	100	100	10	0
August	175	171	150	20	8
Sept.	250	124	190	30	14
Oct.	330	290	133	57	50
Nov.	350	403	298	89	19
Dec.	380	327	223	104	0
1945					
Jan.	430	434	289	145	39
Feb.	350	398	273	125	19
March	350	301	148	153	19
April	250	?	70	47	3
May		?	?	?	?

Jagdpanzer **38** (table title)

* The figures in this chart show the planned production results for every month. They were always set by the Army Weapons Office in the preceding month.

A close look at damage to the building.

A *Jangpanzer* 38 body damaged by bombs.

Service

In April 1944 plans were already submitted that called for the equipping of a *Panzerjäger* company with 14 *Jagdpanzer* 38 for combat testing. Because of the many small problems with the vehicles, there were delays of four to five weeks. Thus, the first 20 *Jagdpanzer* 38 that were finished in April 1944 could not be delivered to the Army Equipment Office in Breslau before May 28-30. Fourteen of them were immediately released for Wa Prüf 6 (2 to Hillersleben, 2 to Bergen, 1 to Wünstorf, 5 to Kummersdorf, 3 to Berka, and 1 to Putlos). With these vehicles, firing tests and cold weather testing were carried out; they also served to establish service instructions and technical advice. The next seven *Jagdpanzer* 38 went to the *Panzerjäger* School in Mielau.

From June 20 to July 25, 1944, another 38 tanks followed, and were used for training purposes by the Replacement Army.

In the period from July 4 to 13, 1944—three months after production began—Army *Panzerjäger* unit 731 finally became the first combat unit to have 45 *Jagdpanzer* 38 assigned to them. It was assigned to Army Group North on the eastern front.

The second unit to receive the *Jagdpanzer* 38, the Army *Panzerjäger* Unit 743, received 45 of them in the period from July 19 to 28, 1944. On August 4, 1944, it was sent on the march to Army Group Center on the eastern front. These two units of three companies each received, according to K.St.N. 1149, 14 light *Panzerjäger* 38 per company. In addition, three light *Panzerjäger* 38 were assigned to the unit staff. In every company there was one, and in the staff company two, *Jagdpanzer* 38 equipped with the long-range Fu 8 radio set. In addition, the *Panzerjäger* unit 731 received four *Bergepanzer* 38 for retrieving and towing broken down *Jangpanzer* 38.

Bergepanzer 38, made in September 1944, without winch and rear spur.

Recovery Tank 38 from January 1945 production, version with winch and rear spur.

Jagdpanzer 38 were assigned to only three other army *Panzerjäger* units: the 741st in September 1944, the 561st in February 1945, and the 744th in March 1945. Unit 741 was divided; its 1st Company went to the eastern front, while the rest of the unit transferred to the Arnheim area, in the Netherlands, as of September 2, 1944.

The *Jagdpanzer* 38 was originally not intended for the independent Army *Panzerjäger* units. Instead, each infantry division was to receive a highly mobile antitank element. *Jagdpanzer* were to be used primarily for counterattacks in enemy breakthroughs, and for the infantry's offensive support.

Thus, the majority of the *Jangpanzer* 38 were assigned to the antitank companies of the infantry, *Jäger*, grenadier, cavalry, and people's grenadier divisions. From August 1944 to January 1945, every *Panzerjäger* company received 14 *Jagdpanzer* 38. From February to April 1945 only ten more *Jagdpanzer* were issued per company, so as to be able to equip a greater number of units.

At the beginning of the Ardennes offensive, 18 *Panzerjäger* companies, plus Army *Panzerjäger* unit 741, were ready with a total of 295 *Jagdpanzer* 38. The Army Group B reported on December 30, 1944, that they had 131 vehicles ready for action out of a total of 190 *Jagdpanzer*. They were divided among 16 *Panzerjäger* companies.

Army Group G reported having 38 of a total of 67 *Jagdpanzer* ready for action. These were with two *Panzerjäger* companies and Army *Panzerjäger* Unit 741.

In all, 67% of the light *Jagdpanzer* were ready for service, and only 38 vehicles were out of action. This shows the value of these vehicles for the German *Wehrmacht*, especially in view of the overwhelming superiority of the Allied forces.

Because of the many interruptions and delays in the production of other armored vehicles, the following units were also equipped with *Jagdpanzer* 38:

- 16th SS *Panzergrenadier* Division "Götz von Berlichingen" (instead of *Jagdpanzer* IV)
- *Panzerjäger* Units *Jüterbog* and *Schlesien* (instead of Panzer IV/70 (V))
- Panzer Division and *Panzergrenadier* Division "*Feldherrnhalle*" (instead of Panzer IV/70 (V))
- *Sturmgeschütz* Brigade 236 (instead of *Sturmgeschütz* III)

On January 24, 1945, the *Wehrmacht* began for the first time to establish a testing unit, which was designated *Panzer-Jagd-Brigade* 104. It was divided into the following units:

- Staff *Panzer-Jagd-Brigade 104* (formerly Panzer-Brigade 104)
- *Panzeraufklärungs-Kompanie* "Krampnitz"
- *Panzer-Jagd-Abteilung* 1
- *Panzer-Jagd-Abteilung* 2
- *Panzer-Jagd-Abteilung* 3
- *Panzer-Jagd-Abteilung* 4
- *Panzer-Jagd-Abteilung* 5
- *Panzer-Jagd-Abteilung* 6
- *Sturmgeschütz*-Lehr-Brigade 111
- *Panzeraufklärungs-Abteilung* 115
- *Panzeraufklärungs-Abteilung* "München"

Every *Panzer-Jagd* unit was to consist of two *Panzerjäger* companies with 14 *Jagdpanzer* 38 each. The reconnaissance companies were to have 16 *Schützenpanzer* (one Sd. Kfz. 251/3, five Sd. Kfz. 250/1, five Sd. Kfz. 250/3, and five Sd. Kfz. 251/21).

The *Panzer-Jagd-Abteilung* 1 also received three, instead of the two planned *Panzerjäger* companies, in addition to their reconnaissance company. Two used the *Sturmgeschütz* IV, and one the *Jagdpanzer* 38. The *Panzerjäger* companies, which were needed for the establishment of these units, were taken from the 21st, 129th, 203rd, 542nd, 547th, and 551st Infantry Divisions, or the personnel units (school companies) 6a, 6b, and 9b, which had not yet been given unit designations.

The *Panzer-Jagd-Brigade* 104 was assigned to the Army Group Vistula on the eastern front at the end of January and beginning of February 1945. Instead of the former tasks of infantry support or counterattacks, the brigade was now instructed to locate Soviet armored units and destroy them with their antitank elements. All in all, the *Panzer-Jagd-Brigade* 104 would have been in a position to inflict heavy losses on the Russians. As was usually the case, the individual parts of the brigade were spread over half the front sector and called on as "firefighters."

On March 15, 1945, there were 51 *Panzerjäger* companies on the eastern front, with a reported combat strength of 359 *Jagdpanzer*

38 (out of a total number of 529). The 26 *Panzerjäger* companies on the western front reported 137 *Jagdpanzer* 38 ready for service out of a total of 236. The four *Panzerjäger* companies in Italy reported a total of 56, of which 40 were ready for action.

The last coherent war strength report before the end of the war, issued on April 10, 1945, listed the following numbers of *Jagdpanzer* 38:

- Eastern Front: 489 ready for action, total 661
- Western Front: 79 ready for action, total 101
- Italy   64 ready for action, total 76

These combined strength reports do not make any claim to being complete, since many units no longer reported. But they show a relatively high percentage of *Jagdpanzer* 38 ready for service. This affirms the fact that these vehicles were mechanically reliable and simple to maintain in service.

Assigned *Jagdpanzer* 38

Month	Number	Unit	Front
1944			
July	45	H.Pz.Jg.Abt. 731	Eastern
	45	H.Pz.Jg.Abt. 743	Eastern
August	14 each	15th, 20th SS, 76th & 335th Inf. Div.	Eastern
	14 each	79th & 257th Inf. Div.	Western
	29	8th SS Cav. Div.	Eastern
	14	97th Jg. Div.	Eastern
Sept.	14 each	306th & 376th Inf.Div.	Eastern
	14 each	83rd, 246th & 363rd Volks Gren. Div.	Western
	14	Kp./H./Pz.Jg.Abt. 741	Eastern
	31	H.Pz.Jg.Abt. 741	Western
	10	16th SS Pz.Gren.Div.	Southwestern
October	14	15thSS & 304th Inf.Div.	Eastern
	14	181st Inf.Div.	Southeastern
	14	349th Volks Gren.Div.	Eastern
	14 each	18th, 272nd, 277th & 708th VGD	Western
	21	22nd SS Cav.Div.	Eastern
	14	44th Gren.Div.	Eastern
	14	4th Gebirgs Div.	Eastern

Month	Number	Unit	Front
Nov.	14 each	344th & 19th SS Inf.Div.	Eastern
	14	337th Volks Gren.Div.	Eastern
	14 each	243rd, 346th, 711th and 716th Inf.Div.	Western
	14 each	9th, 26th, 47th, 62nd, 167th, 326th, 340th & 352nd Volks Gren.Div.	Western
	14 each	z.Jg.Kp.Bock, Lang & Pankow	Western
	10	Supply for H.Pz.Jg.Abt. 731	Eastern
	10	Supply for H.Pz.Jg.Abt. 741	Western
Dec.	14 each	31st SS & 68th Inf.Div.	Eastern
	14	245th Inf.Div.	Western
	7	94th Inf.Div.	Southwestern
	14 each	16th & 79th Volks Gren.Div.	Western
	14 each	252nd, 271st & 320th Volks Gren.Div.	Eastern
	4 each	183rd & 146th Volks Gren.Div.	Western
	4	44th Gren.Div.	Eastern
	25	16th SS Pz.Gren.Div.	Southwestern
	14	Personal Einheit 2a	Eastern
	14	Personal Einheit 33a	Western
	55	Supply	Eastern
	50	Hungary	Eastern
	20	Rebuilt as flamethrowing tanks	Western

Assigned *Jagdpanzer* 38

Month	Number	Unit	Front
1945			
January	14 each	14th SS, 21st, 73rd, 83rd, 129th, 203rd, 211th, 271st, 359th & 384th Inf.Div.	Eastern
	14 each	65th, 278th & 334th Inf.Div.	Southwestern
	14	257th Inf.Div.	Western
	10	181st Inf.Div.	Southeastern
	4	716th Inf.Div.	Western
	14 each	542nd, 547th & 551st Volks Gren.Div.	Eastern
	4	16th SS Pz.Gren.Div.	Southwestern
	14	Personal Einheit 7c	Western
	10	Supply	Western
	26	Supply	Eastern
	14	Pz.Jg.Kp.x.b.V. Pz.AOKI	Eastern
	14 each	2nd & 4rd Kp./Pz.Jg.Abt. 510	Eastern
	25	Hungary	
February	10 each	20th SS, 275th & 600th Inf.Div.	Eastern
	14	356th Inf.Div.	Eastern
	10	Brigade Wirth, Polizei	Eastern
	14	9th Fallschirm Jg.Div.	Eastern
	14	SS Pz.Kp. Saarow	Eastern
	10	Begleit Kp. Reichsführer SS	Western
	14 each	Personal Einheit 6a, 6b, 9a, 9b, 13a & 43b	Eastern
	20	H.Pz.Jg.Abt. 561	Eastern
	31	H/Pz.Jg.Abt. 742	Eastern
	21 each	Pz.Jg.Abt. Jüterbog & Schlesien	Eastern
	8	Supply	Eastern
March	10 each	17th, 71st, 106th, 169th, 251st, 281st, 304th, 305th & 362nd Inf.Div.	Eastern
	10	163rd Inf.Div.	Denmark
	10 each	189th Inf.Div., 716th Volks Gren.Div.	Western
	10 each	6th & 553rd Volks Gren.Div.	Eastern
	10	1st Gebirgs Div.	Southeast
	6	31st SS Inf.Div.	Eastern
	10	Begleit Btl.Reichsführer SS	Western
	102	Supply	Western
	60	Supply	Eastern
	10 each	Personal Einheit Milowitz F & G	Eastern
	21	Pz.Gren.Div. Feldherrnhalle	Eastern
	20	Pz.Div. Feldherrnhalle	Eastern
	31	H.Pz.Jg.Abt. 744	Eastern
	31	StuG Brig. 236	Eastern
April	10	85th Inf.Div.	Western
	10	715th Inf.Div.	Eastern
	4	212th Inf.Div.	Western
	10	38th SS "Nibelungen"	Western
	20	Supply	Eastern
	10	Inf.Div. "Scharnhorst" & "Ulrich von Hutten"	Western
	21	Pz.Jagd Abt. 3	Western
	10 each	Jagd Pz.Kp. 1235, 1245 & 1257	Western
	10 each	R.A.D. Div. Z.b.V. 1, 2 & 3	Western
	10 each	1st & 2nd Marine Div.	Western

Outside Germany

The German Reich sold military supplies to support its allies and pay for deliveries of raw materials from outside Germany. These included armored combat vehicles, including the *Jagdpanzer* 38. In July and August 1944, 15 *Jagdpanzer* 38 were to be delivered to Romania. But since production was not even enough for the German troops, the Romanians received not one single *Jagdpanzer*.

In September 1944 it was decided to make *Jagdpanzer* 38 available to the Hungarian forces. Despite delays, a total of 75 *Jagdpanzer* 38 were delivered to Hungary by rail. 25 of them that were sent on December 7, 1944, arrived on 9 December; 25 more (sent on December 10, 1944) arrived in Hungary on December 12, 1944. The remaining 25 *Jagdpanzer* 38 finally arrived in Hungary on January 13, 1945. The vehicles were assigned to the Hungarian assault gun units, which were fighting with Army Group South on the eastern front.

Experience Reports

The experiences of the first units that were equipped with the *Jagdpanzer* 38 are shown in the following report of October 1944 from the "*Nachrichtenblatt der Panzertruppe.*"

The standpoint of the General of the Armored Troops West affords a look deep into the events of those days:

Abschrift
General der Panzertruppen West
Ia/Pz.Jg.Nr. 1475/44 g.Kdos. 25. November 1944
Abschrift: 7 *Ausfertigungen*
6. Ausfertigung

Standpoint of the General of the Armored Troops West to the report of the Staff *Schanze* included in the enclosure as to *Panzerjäger* (Stu.Gesch.)-Kp. 1708

This is the second case in 14 days in which a *Panzerjäger* (Assault Gun) company coming from refreshing has been wiped out in a few days. To what extent insufficient training of the company, brought on by the shortness of available time and lack of fuel, is to blame cannot be determined from here.

A share of the blame belongs in any case to the **tactical leaders** who, without regard for the uniqueness of the weapon and without considering the suggestions or justified objections of the *Panzzerjäger* company commander or company chief, ordered the action. **Subordination** to an infantry battalion conceals from the start the danger that impossible actions were ordered. **Cooperation** of the assault guns with the infantry unit is to be advocated.

The ordered action to defend the locality in the night of November 14-15 is to be declined **most strongly**. Because of the meager traversing range of the gun, the *Jagdpanzer's* motor must be running to be ready for action. The tank betrays itself, its own crew hears nothing. Especially for a newly established unit with new equipment, sufficient technical maintenance is to be given special care. Small omissions develop in a short time into severe damage, which often bring on premature total losses.

For the action on November 15, the *Panzerjäger* company was placed too far forward. Had the enemy, as usual, begun with preparatory artillery fire, there probably would have been losses of *Jagdpanzer* even before the attack **began**. Had the company been made ready in the midst of the HKF, then it probably could have withstood the reported 14 enemy tanks successfully.

During the transferring movement the *Jagdpanzer* is helpless against enemy tanks and antitank weapons, and thus requires special fire support (quick relocation of thir own *Panzerjäger* platoons, fire support from heavy weapons, and antitank protection from infantry).

The *Panzerjäger* company is the division commander's strongest weapon against enemy tanks, and along with the infantry escort the best weapon for a counterattack. Prerequisites for successful action are:

1. Action in favorable terrain
2. Sufficient terrain scouting, for which time must be allowed.
3. Time for careful technical maintenance of the costly device.

To avoid future such squandering of hard-to-replace materials, it will be regarded as necessary to hold the leaders responsible for this action responsible.

Signed Stumpff

Generalkommando XXXXV.A.K.
Ia Nr. 123/44 g/Kdos. (Stopak)   **3. 12. 1944**
Vorstehende Abschrift mit Anlage zur Beachtung übersandt
Für das Generalkommando
Der Chef des Generalstabes
1 Anlage
Nr. 725/44 g/Kdos. 4. *Ausfertigung*
6. *Ausfertigung der Abschrift*

To the General of the Armored Troops West

1. *Oberleutnant* G ün t h e r, Chief of *Panzerjäger* Assault-Gun Company 1708 3./
Panzerjäger Unit 708. V.G.D.*) has reported today:

2. The company was assigned to the 708th V.G.D. by rail, and unloaded early on November 13 in Rothen (south of Schirmeck). The company had been retrained from the 8.8 cm Pak Mot.Z. to *Jagdpanzer* 38, and had 24 *Jagdpanzer* 38 and an escort platoon of one officer, 4 NCO, and 55 men, all armed with machine pistols. The leader reported to the division while the company unloaded and immediately, at about 5:00 P.M. on November 13, received the order to take action. The immediate action without reconnaissance, without any knowledge of the terrain, and without cooperation with the grenadiers was required, despite the objection of the commander of the *Panzerjäger* unit and the company chief of 1a. For the attack, the company was subordinated to the commander of an infantry battalion (very weak battalion of scattered soldiers gathered under a strange commander). Early on November 14 the planned attack could not be begun, because the authorized infantry support was not in place at the right time. About 1:30 P.M. the Americans attacked with superior forces, and could just be held back by the intervention of the *Jagdpanzer* 38. The request of the company chief to move the guns back a few kilometers for ammunition and fuel supplying and technical checking in the night of November 14-15 was not authorized. The assault guns were to take over the defense of the place at night. Only very weak infantry support was available.

On November 15 at 7:00 A.M. the attack was to be taken up anew, whereby one assault battalion was to take part on the left. Since the assault battalion was not in place at the right time, the attack was delayed for the time being. At 9:15 A.M. the Americans began to attack, after blinding the German HKL with fog.

When the fog dispersed, the Americans were facing the assault guns some 40 to 59 meters away, especially on the left flank of the company. The *Jagdpanzer* immediately took up the firefight. Two *Jagdpanzer* were soon shot down with close-combat weapons after their optics had been shattered, and the two commanders had died of head wounds. At the same time, some six Sherman tanks and enemy infantry, hidden by clumps of bushes, attacked from the left flank. Shortly thereafter, eight Shermans attacked frontally against the company. Another *Jagdpanzer* 38 was destroyed by enemy close-combat weapons. A stuck platoon leader's vehicle had to be blown up. The Americans also attacked with infantry and antitank guns from the right flank shortly thereafter. The battalion leader ordered the relocation into a patch of woods to the rear in a holding action. Since one *Jagdpanzer* 38 had gotten stuck, the *Panzerjäger* company could not follow at once.

The German infantry escort tried to turn back the advance of the enemy infantry and had heavy losses. The company, once the stuck *Jagdpanzer* had been made mobile, set out when the enemy was some 70-80 meters away, and reached the woods; the German infantry was no longer in the woods. The *Panzerjäger* company had only what remained of their escort platoon around them; turning to the right was impossible, because a German minefield was there. In back of the company was a brook that the *Jagdpanzer* could not cross. The company chief decided on a breakthrough to the left toward the village of X. Shortly after coming out of the woods toward X, one *Jagdpanzer* took a direct hit, the company chief's *Jagdpanzer* was hit in the drive wheel, a third *Jagdpanzer* took a direct hit, and all three *Jagdpanzer* burned at once. The last two vehicles also took direct hits from the flank shortly thereafter.

The company lost two officers, missed ten NCOs and men dead, including three commanders, seven men wounded, from the *Jagdpanzer* crews some two officers, thirty men were lost from the infantry escort, nine *Jagdpanzer* were total losses.

3. The commander of *Panzerjäger* Unit 1708 had taken part in the combat. A military court action is to be taken against the company chief.

4. _____ (not legible)

5. The present state of the *Panzerjäger* Assault Unit Company 1708: 5 *Jagdpanzer* 38 in need of long-term repairs (2 gearboxes, 1 gearing, 2 oil-cooler damage), lack of spare parts. Connection with *Panzerjäger* Regiment 111 is taken up.

Technical training good. I-troops with 25 men and tools, well-trained, on hand, one Rescue Tank 38(t) on hand.

Good condition of wheeled vehicles. One "Maultier" is missing. Infantry escort 1 officer, 2 NCO, 24 men. Company still has nine full crews.

Urgent personnel need: one officer, 2 tank drivers, 9 full crews.

The infantry escort platoon will be filled by the division from its own units.

Signed, S c h a n z e, Colonel

* V.G.D. = *Volksgrenadier*-Division

Nachrichtenblatt der Panzertruppen **Oktober 1944**
Experiences with the light *Panzerjäger* 38
The experiences of the first actions of the light *Panzerjäger* 38 justify the following judgment:

1. The light *Panzerjäger* 38 has come through its baptism of fire. The crews are proud of their vehicles; they themselves, like the infantry, have confidence in them.

The all-round-fire machine gun is mentioned with special praise. Strong weapon effect, little overall height, and favorable shaping have proved its full suitability for its two main tasks:

Fighting enemy tanks, direct support of the infantry in attack and defense. One company has shot down 20 tanks in a short time without any total losses of its own. One unit has destroyed 57 tanks (including two Stalin at 800 meters) and had no total losses of its own from enemy fire in these tank battles.

The same unit reached its destination after a day's march of 160 kilometers without breakdowns.

2. The platoon (4 *Jäger*) must be regarded as the smallest combat unit; only then is an effective concentration of fire and mutual support and protection possible. The closed action of units has always proved to be most practical. It leads most surely and speedily to a decisive success, and thus guarantees that the entire combat strength of the unit can be freed quickly for other tasks, while also gaining time for technical maintenance and repairs.

3. Use under platoon strength is justified only in cases of need, or under special terrain conditions. For these cases, fighting as half a platoon must also be taught, and the half-platoon leader trained for independent action.

4. *Panzerjäger* must not be used alone! Especially for offensive actions of all kinds that lead into enemy forces, grenadiers must necessarily be assigned for direct support and close security.

5. Every action must be prepared in advance! Through thorough discussions of the participating leaders, a battle plan must be determined by which the division of tasks, cooperative work of the weapons, and manner of understanding during the battle are made clear.

6. Place of the leaders:
The platoon leader is essentially with his vehicles. The company leader is essentially with the mass of his vehicles; he leads attacks of more than platoon strength himself. The unit leader is with the mass of his unit, or with the unit in action at a focal point. The fighting position of the unit must be where connection points of communication links guarantee trouble-free connection with the grenadier regiments and the division.

7. March:
On a day's march, extending division and loosening are to be carried out, and air-raid gaps of 100 meters and more are to be held.

Marching by night is, what with present drivers' optics, only possible at a very slow speed as instructed by the commander, or by sending a man out ahead of each vehicle. (Changing of drivers' optics is requested and is being carried out.)

8. Preparation:
Entering a readiness area and moving out of the same (except into combat) should be carried out at twilight or in darkness. (Disguise the sound with your own artillery fire!) Villages and squares near railroad station etc. are to be avoided.

9. Camouflage: On the march and in readiness positions, the best disguise has proved to be the turning of the *Jäger* into a bush, to the extent that this fits in with other plant growth. But the camouflage must be removable in a few seconds, so that observation and firing are possible without hindrance.

10. The *Jäger* must have time for technical maintenance and repair service. After marches and between actions, the opportunity for care of the vehicles must be provided; otherwise, they will line the roads instead of being on hand at the right moment as decisive weapons.

11. In combat against tanks and antitank weapons, the low form allows a truly hunter-pike behavior. This allows the light *Panzerjäger* 38 to withdraw quickly from direct enemy view and heavy fire, and to fall upon the enemy anew at effective shot range with concentrated fire.

The front armor withstands the fire of the Russian 7.62 mm antitank gun. Losses have resulted to date only from shots in the side and rear. Thus, it is especially important to show the enemy only the "strong chest." Necessary movements across the front must be made outside direct enemy fire with armor-piercing weapons.

12. To fight against attacking tanks or massed infantry, the *Panzerjäger* 38 are to be kept ready near the front in at least platoon strength, but outside the main zone of enemy preparatory fire. Eye-catching spots like villages, road intersections, and the like are to be avoided.

13. In the support of immediately made counterattacks, the light *Panzerjäger* 38 has often contributed decisively to a quick victory. It must never be applied alone! The infantry must be informed in

advance, since the *Panzerjäger*, after a successfully carried-out counterthrust, must drop back for ammunition, even during the fight.

14. Communication with grenadiers during combat was maintained through direct calling. Targeting by grenadiers was done by firing flare pistols, or calls to the commanders.

15. The light *Panzerjäger* 38 is not suited for supporting attacks along roads that run through swampy land. It gets stuck when it leaves the road, and on the road it is undefended against enemy antitank guns that are hidden in the terrain, because of the limited traversing field of its cannon, and will be shot down on its lightly armored sides.

16. For the use of the light *Panzerjäger* 38 within fully motorized units or battle groups (for example, for powerful reconnaissance advances), the light *Panzerjäger* 38 is too slow. It is worn out prematurely through such action, and has unnecessary breakdowns through technical damage.

17. In street fighting (in Warsaw) the light *Panzerjäger* 38 has proved itself well through its mobility on the road. The all-round fire of the machine gun has been used here to particular advantage. Since the gunner must open the hatch for reloading, it is necessary that another vehicle should give fire cover. (Advance understanding by radio.)

18. In training, repeated practice must be provided for:
- radio contact,
- the use of the all-round-fire machine gun, and
- cooperation with grenadiers. This applies above all to the infantry, who must be given more instruction in cooperation with tank destroyers and assault guns in their basic training from the start in the replacement army. (Cooperation between training units of the armored troops and the infantry!)

As already noted, the first combat unit to be supplied with the *Jagdpanzer* was Army *Panzerjäger* Unit 731. From July 1944 to the war's end they fought on the eastern front. Originally equipped with 45 *Jagdpanzer* 38, the unit received ten replacement vehicles in November 1944 and twenty more in December. The available reports state the strengths:

Date		Listed	Ready for action
1944	September 1	33	30
	October 1	24	18
	December 1	26	11
1945	January 1	22	12
	February 1	41	27
	March 15	28	13

On January 21, 1945, the commander of Army *Panzerjäger* Unit 731 submitted the following experience report no. 6, relative to the *Jagdpanzer* 38, to his superiors:

The report includes the time from 12/1/1944 to 1/31/1945. The third battle of Courland is included as a combat action in the time period of the report.

The prelude to the Courland battle was a drumfire of about three hours, which extended to the distance of the division's command post, and was intensified by the steady attacks of bombers and fighter planes to the distance of the corps command post. After the H.K.L. was destroyed, the enemy achieved breakthroughs in the sector of the XXXXVIII Armored Corps as far as the regimental command posts. He could not accomplish deeper penetration with operative action possibility. With his armored weapons, the Russian followed his infantry forces' breakthroughs only hesitantly, and in wolfpacks.

The tank defense supported itself, after the loss of antitank guns to the drumfire, excellently with assault guns, light *Panzerjäger*, and self-propelled guns. About the action of Army *Panzerjäger* Unit 731 it is to be reported that it stood ready at the beginning of the battle with masses of men as the corps reserve, and with one company as the division reserve of the 205th Infantry Division. After marking off the breakthrough area on the corps' sector, the two remaining companies were each subordinated to a division. On the first day of the battle, each company had seven *Jäger*.

The sudden onset of frosty weather brought on heightened technical difficulties (especially breaking of bevel drives and gearings), so that only 60% action strength was attained for the day. The battlefield was mainly flat, crossed by several high ridges that were fought hard for, cut by low areas and ditches, decked with woods and swamps whose upper layers were frozen. These terrain conditions limited the action of platoons to four *Jäger*. The drivers of the grenadier regiment paid attention to the action suggestions of the company chief, and also urged the formation of even smaller groups. The enemy put up a tough fight, but his infantry and armored weapons showed clear command difficulties. His strongest factor was always the concentrated fire of his heavy weapons. In the course of the companies' action, some experiences stood out that emphasized experience gained to some extent previously. For the choice of an assembly area it is important to find a space from which good marching routes lead to the regimental sectors of the divisions. It must be outside the immediate area of heavy enemy fire, for otherwise there would be heavy losses from artillery and grenade launchers. A dependably sure communication connection with the division or unit is also needed.

On that day the impulsive immediate counterattack with little covering infantry fire proved itself. For carrying out several counterthrusts at the same object, various approach routes must be chosen. For example, the 3rd Company had to lead three different counterthrusts in one day against a village on a commanding height. Only by seeking and finding a new approach route for the

Jäger again and again could a new moment of surprise always be found, and thus success could be assured.

Night attacks were particularly successful. But they cannot last until daylight. The advantage of night attacks is that the enemy antitank defense is remarkably limited in finding sure targets, and can be overcome more easily. The farther one's own counterattack goes on successfully, the greater is the danger of encountering heavy weapons, particularly the enemy's antitank guns.

If some tanks find themselves at dawn in an enemy defensive position, they will be surprised by the enemy weapons, since spotting enemy defensive weapons was impossible at night. In such situations the company lost three *Jagdpanzer*.

The following experience was gained in carrying out night attacks:

1. The approach route cannot be taken under aimed and observed enemy fire in the dark in country otherwise observed by the enemy. Firing heavy enemy weapons in the direction of a sound is usually ineffective.

2. The enemy tank defenses are strongly hindered in their targeting, and can be overrun easily.

3. The approach of track noise has an extraordinarily strong effect on the enemy's morale at night.

4. The driver has very limited vision, and must constantly be directed by the commander.

5. The commander must stand in his hatch and observe, even under heavy fire. The shear scope and periscope cannot be used. This leads to very high losses of officers and commanders, unrelated to other losses.

6. The precise recognition and aiming at a target with the aimer's aiming scope is possible only in bright moonlight. The target thus can only be aimed at crudely by the commander. Fire must be kept high if your own infantry is passing by the *Jagdpanzer*, so as not to endanger your own troops. Thus many shots go too far.

7. Firing the radioman's all-round-fire machine gun can be done successfully only when standing up in the hatch.

8. Concentrating the fire of all the *Jagdpanzer* by the tactical leader is not possible as a rule, for seeing from one *Jäger* to another is difficult. The *Jagdpanzer* are thus definitely fighting alone in the darkness.

9. Cooperation with the accompanying infantry is lost because of the poor vision conditions. It is often lost under heavy enemy fire and cannot be reestablished, since the commanders are thoroughly busy observing the battlefield. Thus, the *Jagdpanzer* is particularly endangered by close enemy combat. The commandant must always have his machine pistol and hand grenades within reach.

10. When tracer bullets are fired, your own *Jagdpanzer* offers a target visible from afar.

11. Accompanying infantry should hesitate to occupy a trench that has been occupied by the enemy.

12. The danger of getting your vehicle stuck increases with bad vision conditions.

The following battle reports by two companies illuminate individual phases of combat in the course of the battle.

On the second day of the battle, the 3rd Company received the order to take four combat-ready *Jäger* and retake the village of Balki on a commanding height.

Under cover of a trench, the company could move to 50 meters behind the position of the infantrymen who were ordered to escort it. A patch of bushes allowed them to reach the starting position of the assault unseen by the enemy. There the company leader discussed the expected course of the counterattack with the platoon leader of the accompanying grenadiers.

The enemy's main resistance was expected in the trench sections just northeast of Balki. The southern part was occupied by two known machine-gun positions. Since the counterattack could not be split up, it had to be directed first at the trench section just northeast of Balki. Four infantrymen were assigned to the *Jägers* for close defense. The infantry (15 men) moved up close behind a row of bushes just before Balki, and were to attack the enemy trench on a white flare signal from the command *Jäger*. The four *Jägers* would attack the enemy from the flanks.

The enemy was ready to flee, and was attacked effectively with explosive grenades. During this burst of fire, the infantry attacked and occupied the trench.

Nests of opposition were taken in close combat. Through the successful counterattack of the small battle group, a dominant height was again in German hands. The counterattack against the height had to be repeated twice during the course of the day. In a firefight between four *Jägers* of the same company and a "J.S. 12" tank at a range of 1200 meters, it was seen that ten aimed shots from the enemy tank at the chief's *Jäger*, which was in a well-hidden position, were well aimed, but always fell some 100 meters too short. The company chief immediately had one *Jäger* go off to the right and attack from the flank through a hollow. With this *Jäger's* sixth shot, a broadside hit, the "J.S. 12" burned out. This emphasizes the experience that when possible, one *Jäger* alone should never carry on the firefight, for when firing, the powder smoke got into the shear scope and hindered the commander from observing and correcting the firing. A second *Jagapanzer* could give the shot corrections by radio that led to the destruction of the enemy tank.

The 2ⁿᵈ Company reports on carrying out a counterattack:

1.Picture of the enemy:
The enemy had established himself at the point of woods northeast of Dadzi in double company to battalion strength, and remained generally quiet until darkness set in.

2.Terrain:
Slightly rolling, spotted at this place by groups of bushes and small patches of woods, with the ground frozen hard.

3.Task:
Assembling the company within the framework of Major Singrün's barrage company at the battalion command post H.Pi.Btl. 630, near Skuenieki.

4.Course of the Battle:
About 6:00 P.M. a surprising enemy breakthrough with about sixty men takes place in a patch of woods. Here the enemy can occupy a light howitzer battery. Through a counterattack carried out at once, and without heavy weapons, the enemy was thrown back to his initial position to loud cheering. The company's three *Jagdpanzer*, meanwhile alarmed, take up firing positions right behind their own HKL about 6:40 P.M. About 8:00 P.M. a surprising enemy breakthrough with about the same strength takes place for the second time. Here, too, the enemy is able to occupy the battery and push forward to the level of the battalion command post of our neighbor on the right.

Through a frontal counterattack, again immediately undertaken, the enemy is thrown back. Through well aimed fire from the cannons and machine guns of the three *Jagdpanzer*, the penetrating enemy is struck on his right flank and pulls back as if in flight. The enemy leaves a few dead men on the battlefield and a few prisoners in our hands.

5.Experiences:
In night action, precise scouting of the terrain, the approach routes, the firing positions, and one's own H.K.L. is absolutely necessary. Make contact with the infantry leaders in the occupied sector (if possible, also neighbors). Short-range securing of the *Jagdpanzer* by a few grenadiers is vital. The aiming gunner must aim at the previously chosen target in the light of the tracer bullets. Strictly conducted fire control. When firing over one's own infantry, firing boundaries must be set.

In tank warning and recognition service it is advantageous to detail a warning sentry in every situation—no matter whether in assembling or under heavy drumfire. This man is simultaneously an airplane spotter, and gives warnings if enemy planes approach.

6.Driving on ice:
The lack of ice tunnels has proved to be a disadvantageous. The *Jagdpanzer* will slide off sideways or skid.

7.Tank recovery:
It is practical to bring the advanced maintenance support point with the rescue troop as close as possible to the combat area.

In conclusion, it could be determined again in the combat days that the *Jagdpanzer* 38 did its job very well as a fast, mobile *Panzerjäger* and hard-shooting support weapon of the infantry in attacks and counterattacks. On the first day of the battle the following results were scored in the course of two hours:

Destroyed:	1 "J.S. 122" tank
	1 T 34 tank
	3 7.62 antitank guns
	8 machine guns
	3 grenade launchers
	60 enemy dead
Captured:	2 prisoners
	4 machine guns

I.V.
Captain

Distribute:
Inspector General of the Armored Troops, 1 x direct
"   "   "   **1 x a.d.D.**
-2 enclosures (rescue report, technical experience)

Technical Experience with the *Jagdpanzer* 38

1. Running Gear:
The eye of the spring ram has broken out by the seat of the wedge in several cases.

Suggestion:
Strengthen the eye of the spring jack. This will result in an enlargement of the hanging apparatus.

On the new-type leading wheels (pressed), the rim of the leading wheel has often been bent.

Suggestion:
Strengthen the leading wheel by putting in ribs.

2. Track Bolts:
It has been ascertained several times that the old-type track bolts break at various places as a result of becoming nicked.

3. Track tightening:
The bond of the ratchet key to tighten the tracks is very weakened. The corners of the square break off.

Suggestion:
Strengthen the bond.

4. Clutch-Steering brakes:
In the Sfl 38(t) a spring was built into the rod for the steering coupling, allowing it an elasticity while starting. Through the stiffness of the rod, in many cases the uncoupling jack has broken off between the two attachment holes.

Suggestion:
Make the change as with Sfl. 38(t), and strengthen the uncoupling jack.

5. Steering traverse:
The steering traverse is attached to the jack by six hexagonal screws. In two cases, all the screws have been torn off.

Suggestion:
The steering traverse has two machined guiding surfaces on its sides that fit exactly into a matched cutout in the jack. The power is taken up in this arrangement not by the screws, but by the steering traverse and the jack. Adapt as shown in enclosed sketch. (missing)

6. Clutch coupling:
The two halves of the divided clutch coupling are screwed together with six cylinder-head screws. These screws loosen, since they are not secured. Because of faulty attachment, the clutch couplings are torn in several cases.

Suggestion:
Secure the cylinder-head screws according to the enclosed sketch.

7. Flame extinguisher:
The flame extinguisher has proved to be unsuitable. Through it, the motor sound can be heard at great distances. Letting the motor run warm in the assembly area has thus become impossible.

Suggestion:
Reuse the old muffler, which has proved itself well, or install a sound damper in the flame extinguisher.

8. Aprons:
The aprons are fastened by screws that are electrically welded on. The screws often break at the welding point. Repairing at the protruding support points is only possible when an electric welding device is at hand.

Suggestion:
Attach the aprons with screws through holes bored in the side armor. Spread the screws out about 210-15 cm.

9. Anti-skidding profile:
On hard frozen ground and slick roads it is not possible to handle major slopes, since the tracks skid. It is very disadvantageous that the tracks have no anti-skidding profile.

Suggestion:
Equip the vehicles as before with grippers on the treads.

10. Cone drive:
Noticeably great failure on *Jagdpanzer* through damage to cone drives. Thorough investigation gave these results:

a) Tooth play is noticeably great through wear of the ball bearings.

b) One-sided wear pattern [?]. Lack of precision in manufacturing.

c) Material is apparently not tough enough.

d) Drive parts are too weak. The parts were originally made for the old Czech Panzer 38 tank.

Suggestion:
Use cone wheels with Gleason teeth.

Revised *Jagdpanzer* 38

Twenty *Jagdpanzer* 38 were rebuilt as **Flamethrowing** tanks for Armored Flame Companies 352 and 353. Little has been found in the literature to date about the use of these vehicles.

A b s c h r I f t
Oberkommando Heeresgruppe **G.O.U., 1/20/1945**
Pz.Flamm-Kp. 352 G e h e i m Kp.-Gef.St., 1/17/1945
Re.: Experience report on the use of the Flame Tanks
Reference: 25th ***Panzernradier* Division, Unit 1a No. 59/45 geh. Of 1/16/1945**
To the 25th ***Panzergrenadier* Division via Panzer Unit 5**

After the Pz.Flamm-Kp. 352 was first assigned to the 36.VGD, but did not see action there, it was subordinated to the 25th *Panzergrenadier* Division on 1/6/1945. The Pz.Flamm-Kp. 353 was previously subordinated to the 17th SS Division and saw action there.

Flammpanzer 38 (Flamethrowing tank)

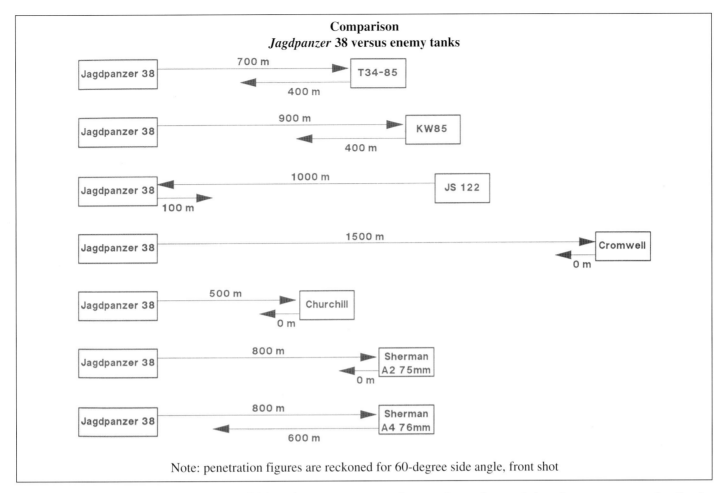

Comparison
Jagdpanzer **38 versus enemy tanks**

Note: penetration figures are reckoned for 60-degree side angle, front shot

As for the procedures and the events of this action, a report should be requested of the 17th SS Division. The remains of Pz.-Flamm-Kp. 353 (4 tanks and the entire supply train) were subordinated to Pz.-Flamm-Kp. 352 on 1/6/1945. After their transfer to the Weissenburg area, five flame tanks were ready for action, and took part in the attack on Hatten on 1/9/1945 in the group of Pz.Abt.5.

In this action the flame tanks proved themselves fully in action against bunkers, field fortifications, and in towns. The prerequisite was the effective support of heavy tanks, since the flame tanks are defenseless against armor-piercing weapons because of their weapons. In action at Rittershofen on 1/13/1945, two flame tanks of the 1st Attack Group were lost to direct hits when they came upon an enemy antitank-gun front. The 2nd Attack Group, after initial success in town fighting, came upon a mine barrage, where one flame tank was lost, and the other was put out of action by close-combat means. In action it was seen that the flame tanks were equal to any difficulties in technical terms. The great loss of flame tanks on the march in the Weissenburg area is attributable to the fact that the vehicles had not been broken in sufficiently, and thus broke down in the difficult road conditions (icing and repeated steep upgrades), with minor motor and gearbox damage. When the vehicles were delivered, they had only traveled 25 to 50 kilometers.

The built-in flame apparatus and the burning oil tanks have functioned faultlessly without significant damage.

The crew of the tank, set up at the unit's establishment, numbered four men (driver, flame-shooter, radioman, and commander), unlike the originally planned three-man crew (driver, flame-gunnter, and radioman, as well as commander), proved to be right and necessary in action.

Greater difficulties have actually occurred only in obtaining spare parts, since neither the 25th Armored Grenadier Division, nor the Panzer Unit 5 were prepared for these special vehicles in terms of workshops and spare parts. When the company was established it was given no spare parts, and those that were sent later turned out to be wrong and unusable.

If enough spare parts are delivered, then the repairs of the broken-down vehicles by the 1.-Panzer *Abteilung* 5 could be just a matter of days.

In conclusion, the following can be said about the action of the flame tanks after the previous combat experience: The flame tanks, on the basis of their equipment and armor, are very good in a situation where they support the grenadiers in action against bunkers, field fortifications, and in towns. Their action has shown that the enemy is very sensitive to the actual and morale effects of the flame tanks. A prerequisite for successful action is support by heavy tanks or heavy infantry weapons to knock out the armor-piercing weapons against which the flame tank is helpless on account of its armament and equipment. It is necessary that the flame tanks are never sent into action singly, but at least in platoon groups, or best of all, in company groups.

Signed, Ritter (Lt. And company Leader)

25th Panzer-Grenadier Division   Div.Gef.St. 1/17/1945
Abt. 1a Nr. 74/45 geh.
Reference: FS Gen.Kdo. LXXXIX.A.K.
**   1a Stopak Nr. 30/45 geh. Of 1/16/1945**
Re: Experience report on the action of flame tanks.
To the
***Generalkommando* LXXXIX.A.K.**

With the enclosed experience report I am in agreement.
The flame tanks have proved themselves fully in action to date.
M.d.F.g.
Signed Burmeister (Major General)
F.d.R.d.A
Captain

Jagdpanzer IV

Development

The extraordinary success that the units equipped with assault guns could report about fighting against tanks, and the wish of the industry to have to complete just one more chassis, led in September 1942 to the order from the Army Weapons Office to create the *Jagdpanzer* IV. The contract went to the Vomag* firm in Plauen.

The carrier of the body was foreseen as a modified Panzer IV chassis, which could carry the 7.5 cm Pak 39 L/48. On May 13, 1943, Adolf Hitler was shown the wooden model of the new body. The temporary design showed an unchanged Panzer IV chassis, which carried a 7.5 cm cannon in a low upper body. This first version had an overall height of only 1700 mm. Despite the truly ideal low body, one soon recognized the boundaries that a hilly landscape could cause for the possible firing height. The angled armor plates and the gun shield of the design, which showed no shell-catchers, unlike the assault gun, could scarcely be improved.

During the summer of 1943 the design was constantly improved. The most important changes concerned the bow of the vehicle. The armor on the front of Panzer IV had grown during the war years from 14.5 to 80 mm.

* Vogtländische Maschinenfabrik AG. (Vomag)
1915- 1932 Vogtländische Maschinenfabrik AG, owned by J.C. and H. Dietrich, Cranach Street 4, Plauen.
1932-1938 Vomag Betriebs-AG.
1938-1945 Vomag Maschinenfabrik AG.
Since 1915 it produced trucks; its monthly capacity was about 85 to 100 vehicles, assembled by some 1200 employees.
In 1932 bankruptcy had to be reported. A receiver company was able to bring the firm back into the black as Vomag Betriebs-AG. Within the "Fast Program," trucks and buses were produced until 1942. Since 1941 it had been converted to the assembly of armored vehicles. In 1945-46 the plant, only slightly damaged in the war, was completely dismantled as war booty.

Wooden model of a *Jagdpanzer* body made by the Vomag firm on the Panzer IV, Type F chassis. The machine-gun ports were not yet planned, and the driver looked through a single periscope.

Jagdpanzer IV - 0 Series

January 1944 - Chassis no. V 2
(Muzzle brake, rounded upper front armor, pistol loophole in the upper left side - MG 42 by the
aimer's seat - Sfl Zf 1a targeting scope - Periscope installation - two MG ball mantlets on the front
of the upper body - periscope at left front - curved guide for vision device in front, closed from
inside).

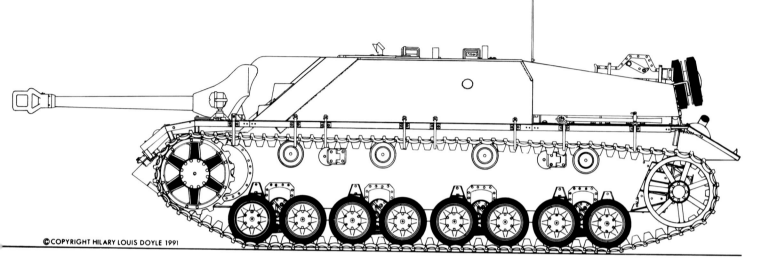

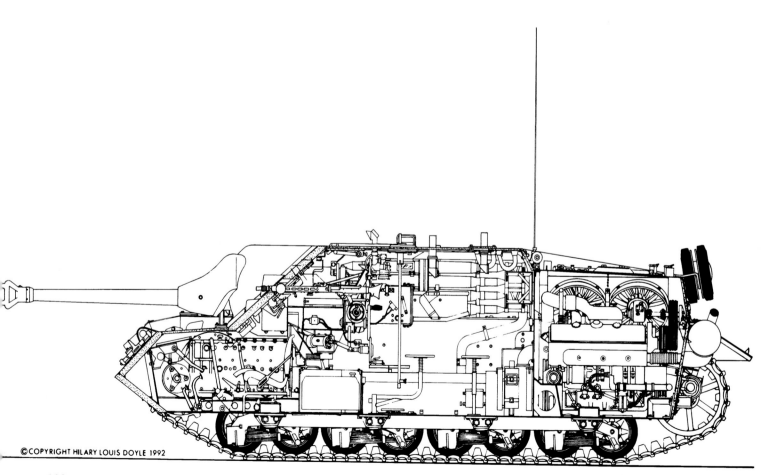

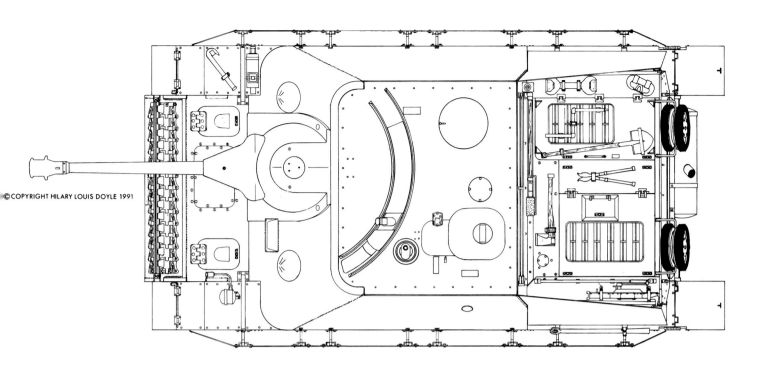

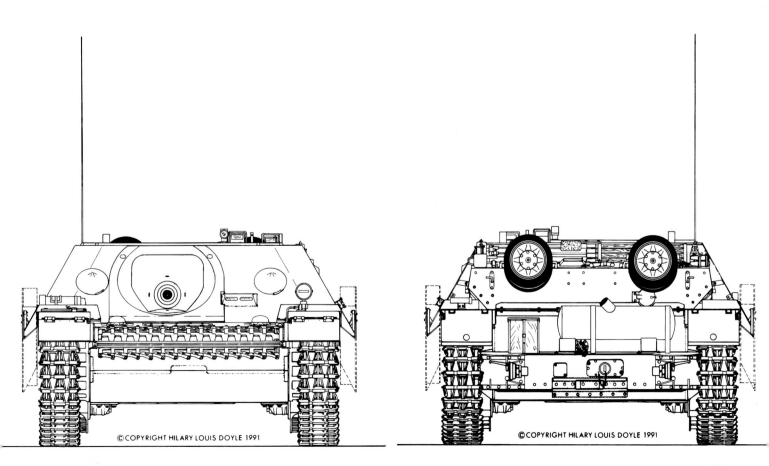

©COPYRIGHT HILARY LOUIS DOYLE 1991

©COPYRIGHT HILARY LOUIS DOYLE 1991

©COPYRIGHT HILARY LOUIS DOYLE 1991

101

The soft steel body of the prototype already shows the final form of the *Jagdpanzer* IV. Before this vehicle was approved for series production the inner and outer peephole armor was reworked again, and a bracket for the left track cover was planned.

The angle of the forward armor plates measured only 15 degrees. While the 80 mm armor was adequate at normal ranges against the 7/5 cm cannon of the U.S. "Sherman" tank and the 7.62 cm gun of the Russian T 34, it gave way to 17-pound British tank shells and the new 8/5 cm long-barreled cannon of the T 34/ 85. In February 1943 the Army Weapons Office had suggested improving the front armor of the Panzer IV by tilting the armor plates more sharply. The suggestions, though, were always rejected, with the comment that such a change would place an unbearable burden on the series production.

When production of a completely new vehicle, the *Jagdpanzer* IV, began, the possibility of a change was provided. The upper plate of the new front armor was 60 mm thick, and set at a 45-degree angle, while the lower plate, 50 mm thick, had an angle of 55 degrees. Thus, it provided protection that equaled a rolled-steel thickness of 110 or 123 mm.

One of the vehicles that belonged to "Test Series *Panzerjäger* 39." It served as a school vehicle. The pistol loophole on the body sides was welded on. The tracks with diagonal ice cleats were found only rarely on the production vehicles, since they made the tracks heavier.

Rear view of the *Jagdpanzer* IV.

The roof of the *Panzerjäger* 39 test series held a third, removable periscope at the commander's left.

The *Jagdpanzer* IV prototype, made of soft steel, was shown to the Reich Chancellor on December 20, 1943. On the upper bow plate, on each side of the primary armament, was a round opening to hold a machine gun. In the prototype the side armor plates curved into the front armor. In production, though, only flat plates were used, so as to save time and cost. Round machine-pistol loopholes were planned for both sides of the body, but they too were omitted in production, since the installation of a "close-range defensive weapon" in the roof was now planned.

Development of the Designations for the *Jagdpanzer* IV

kleiner Panzerjäger und Panzerjager der Fa. Vomag
Führer;s conderence,   5/13/1943
Le. *Panzerjäger* auf Fgst.Pz.Kpf.Wg.IV
"Overview of the Army's Armament State" Chef-H.Rüst.u.BdE/Stab Rüst III.
                                                              7/15 - 8/15/1943
Panzerjäger auf Fahrgestell **Panzer IV**
Führer's conference   8/19/1943
7.5 cm Pz.Jäg.39 (L.48) IV
"State of Development, Chef.H.Ru:st.u.BdE/Wa Prüf 9/15/1943

This vehicle, of the "*Panzerjäger* 39 Test Series" (chassis no. 2), was delivered to the troops. The pistol loopholes in the sides and the third periscope opening in the roof are welded shut. The "close-combat defensive weapon" is not yet installed, and the opening for it in the roof was covered by a round plate. The body shows bullet holes from hollow-charge weapons, made during postwar ballistic testing.

leichter Panzerjäger mit 7.5cm L/48 auf Fahrgestell IV

Führer's conference	9/30/1943

Panzerjäger IV für 7.5cm Pak 39 L/48 (Sf.)(Sd.Kfz. 162)

O.K.H. (Chef H.Rüst u. BdE) In 6	Oct. 1943

le. *Panzerjäger* auf Fgst.Pz.Kpf.Wg.IV mit 7.5 cm Pak 39 L/48

"Overview of the Army's Armament State, Chef.H.Rüst.u.BdE/ Stab Rüst III.	10/15 - 12/15/1943

Sturmgeschütz n.A.

Gen.St.d.H./Org.Abt.	11/13/1943

l. *Panzerjäger* (Fahrgestell IV m. 7/5cm L/48)

Chef.H.Rüst.u.BdE, Wa.Abn.	1 /2- 8/31/1944

le/Pz.Jg.39

Chef.H.Rüst.u.BdE, Wa.Abn.	7/1/1944

Pz.Jg. auf Pz. IV

GenSTdH/General der Artillerie Kriegstagebuch	1/21/1944

Panzerjäger Vomag and *Sturmgeschütz* n.A

Führer's conference	1/25/1944

Sturmgeschütz neue Art

GenSTdH/General der Artillerie Kriegstagebuch	1/26 & 4/16/1944

Instead of "*Sturmgeschütz* neue Art" the designation
le.Pz.Jäger (IV)
Gen Insp d.Pz.Tr. to OKH/Wa Prüf

K/St.N. 1149	2/1/1944

Sturmgeschütz **n.A. auf Fgst.Pz.Kpf.Wg.IV mit 7.5cm Pak 39 L/48 (Sd.Kfz. 162)**

"Overview of the Army's Armament State" Chef.H.Rüst.u.BdE/ Stab Rüst III.	
	2/15 - 4/15/1944

7.5 cm *Panzerjäger* 39

Chef.H.Rüst.u.BdE, Wa.Abn.	2/29 - 9/6/1944
le.Pz.Jg.IV Gen. Insp.d.Pz.Tr.Akten	3 /4 – 10/- /1944
Stu.Gesch.n.A. (Vomag Pz.Jg.)	
Wa Prüf 6	3/28/1944

Note: not to be designated as **Stu.Gesch.43**, (because:
1. Year forbidden by Führer
2. All 43s were 8.8cm caliber to that time!)

"Overview of the Army's Armament State" Chef H. Rüst.u.BdE/ Stab Rüst III.	4/15/1944

Le.Pz.*Jäger* IV 7.5cm Pak 39 L/48

Wa Prüf 6	5/1/1944

Sturmgeschütz **n.A. mit 7.5cm Pak 39 L/48 auf Fgst.Pz.Kp.Wg.IV**

"Overview of the Army's Armament State" Chef.H.Rüst.u.BdE/ Stab Rüst III.	5/15 - 10/15/1944

Naming of *Sturmgeschütz* n.A. is again urged by Org.Abt. GenSTdH/General der

Artillerie Kriegstagebuch	6/4/1944
le/Pz/Jg.39 auf Pz.Kpfw.IV (Vomag)	
Wa Prüf 6	6/2/1944

Sturmgeschütz **n.A. mit 7.5cm Pak 39 L/48**

Wa Prüf	7/5/1944

Pz.Jg.39

s.H.Pz.Jg.Abt 525	9/1/1944

le.Pz.Jg.Vomag mit 7.5cm Pak L/48 auf Fgst.Pz.IV as
"**le. *Panzerjäger* IV**" (ehem. Stu.Gesch.n.A.)

Chef GenSTdH/Org.Abt./Gen.Insp.d.Pz.Tr.	9/8/1944

Called by the troops:
Jagdpanzer IV
Called by regulations:
Jagdpanzer IV Ausf.

GenSTdH/Org.Abt.Gen.Insp.d.Pz.Tr.	9/11/1944

Jagdpanzer **IV Ausf. F**

D653/39 Wa Prüf 6	9/15/1944

Jagdpanzer **IV**

Gen.Insp.d.Pz.Tr.Akten	10/19 - 12/4/1944

Jagdpanzer IV und *Panzerjäger* IV (7.5cm Pak 39 L/48) (Sd.Kfz.162)

K.St.N. 1149	2/1/1944

Jagdpanzer IV, *Panzerjäger* IV (m. 7/5cm Pak 39 L/48 (Sd./ Kfz.162)

"Overview of the Army's Armament State", Chef.H.Rüst.u.BdE/ Stab Rüst. III.	
	11/15/1944-3/15/1945

Jagdpanzer **IV** Verskraft	1945

Specific Features

The basic chassis of the *Panzerkampfwagen* IV, Type F, with running gear and drive aggregates was taken over unchanged for the *Jagdpanzer* IV; so were the Maybach HL 120 TRM motor, the ZF SSG 76 six-speed gearbox, the steering system, and the gearing. The engine compartment with its cover, the rear part of the hull, and all components of the running gear, including drive wheels, return rollers, jack rollers, road wheels, and tracks were also left unchanged. The following parts of the chassis differed from Panzer IV, Type F:

-   The bow of the hull was sharply angled, unlike Panzer IV,
-   The emergency exit now had a rectangular shape and a new latch. The exit formerly by the radioman was moved to a left central position in the hull roof, directly under the aimer's seat.
-   Instead of the DKW aggregates (to turn the turret) of the Panzer IV, the *Jagdpanzer* had an additional fuel tank. While the two main tanks of the Panzer IV were located under the turret stage, they were moved forward under the gun to save space. Thus, the fuel filler cap opening also had to be moved.

- The arrangement of the brake ventilation system, the fighting compartment heater, the radio installation (which differed from the Panzer IV in having a new component, one *Funkgerät* Fu 5 and, in leaders' vehicles, also one *Funkgerät* Fu 8 set) were changed.
- All the equipment was housed differently than in the Panzer IV, corresponding to new space conditions.

A look through the opening on the left side of the upper body, otherwise covered by a conical lid. It allowed shooting with a machine pistol or MG 42.

Series production *Jagdpanzer* IV, left side of the vehicle.

Series production *Jagdpanzer* IV, seen from the right front.

The series production of the *Jagdpanzer* IV with the 60 mm front armor shows the straight shape of the upper front panel. Made in the first months of 1944, the vehicle was sent to a training unit. Fourteen spare track links were carried on the upper front bow plate. Two spare road wheels were in brackets on the rear plate.

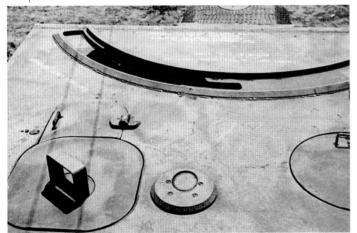

The roof of the *Jagdpanzer* IV fighting compartment. The cover of the sight channel is missing. This *Jagdpanzer* IV is one of the few early models that were fitted with a "close combat defensive weapon."

The experimental installation of an all-round-fire MG 34 on an early *Jagdpanzer* IV.

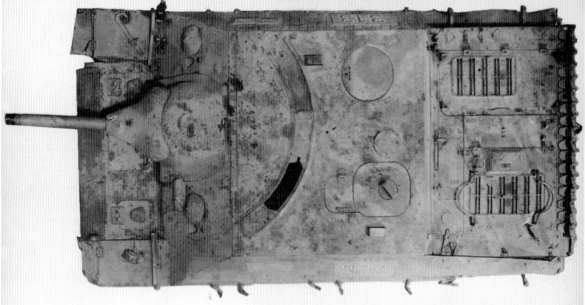

Facing pages: A *Jagdpanzer* IV of the Panzer Instruction Division was given a thorough inspection in Britain. The muzzle brake was removed by the crew. The spare track links were now in back on the rear wall of the engine compartment. Two spare road wheels were on the engine compartment flap for the radiator. The length of the curved sight channel was shortened by a welded-on panel. The dark yellow basic color was enhanced with camouflage paste by the crew.

While the hull remained almost unchanged, the *Jagdpanzer* IV received a completely new upper body. The body was closed on all sides; the 7.5 cm Pak 39 (L/48) was the primary weapon. All the body panels were angled as well as possible. The bow plate (60 mm thick) tilted 50 degrees, the 40 mm thick side plates 30 degrees, and the rear 30 mm late inclined 33 degrees.

The gun was cardan-mounted in a mantlet that was attached to the front armor plate by wedges. It was 200 mm right of center. The weapon mount consisted of a ball-shaped holder inside; outside a "pig's head" shield protected the cannon.

The crew received the following auxiliary weapons:

- one MG 42 machine gun,
- MP 44 or MP 40 machine pistols.
- one close-range defensive weapon for armored vehicles, on the roof.

This close-combat weapon for use against infantry could be turned 360 degrees, and fired either fog cartridges or 2.6 cm explosive shells. There was room in the vehicle for 79 rounds of cannon ammunition, plus 1200 MG and 384 MP rounds, and 16 rounds for the close defense weapon. As a rule, the *Jagdpanzer* carried 50% explosive shells and 50% tank grenades. This ratio could change according to action.

The crew of the standard vehicle consisted of four men:

- the gun leader, simultaneously the tank commander,
- the aiming gunner,
- the loading gunner, and
- the driver.

The decision was made that the shooting opening by the driver gave no advantages. Until production could be changed, the opening was welded shut by means of a conical armor plate. This vehicle (chassis number 320106) was destroyed by a shot that penetrated near the driver's sight opening.

The crew of the leader's vehicle also included a radioman.

For vision, the driver had two periscopes (prismatic scopes) for his use (instead of the glass blocks used on the Panzer IV). The aiming gunner, who sat right behind the driver, saw through a self-propelled-mount aiming telescope (Sfl ZF 1a) and a periscope. The opening for the aiming scope in the roof was protected by a movable cover slide. The tank commander, who sat right behind the aiming gunner, had a visor with a divided lid. The front part of the lid could be opened to set the shear telescope in it (Sf 14 Z). For observation, the commander also had one turning and one fixed periscope to use.

The loading gunner was also the radioman, and observed to the right with a periscope, as well as through the visor by the machine gun in front. Over his battle station he could open a round hatch in the roof.

As with all other German armored vehicles that were produced over a long period of time, one could not speak of a single "standard" version of the *Jagdpanzer* IV either. Suggestions for changes flowed constantly into production, simplified the assembly, and improved the quality. Problems in supplying also required some changes.

During the production of the *Jagdpanzer* IV the following noteworthy changes were introduced:

Jagdpanzer IV (chassis no. 320106); view of the fighting compartment from above. Some fuel tanks were moved forward. The rectangular emergency exit hatch is in the middle of the compartment.

Pieces of the "pig's head" shield.

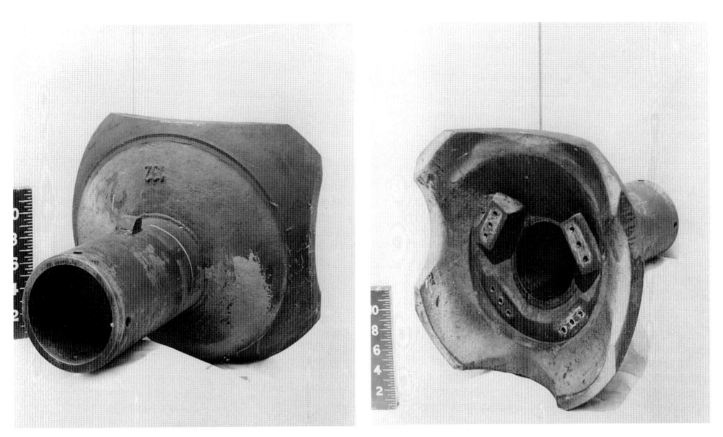

Details of the inner mantlet armor.

Turning frame with shield pins for the cannon.

Mount for the ball mantlet.

Details of the recoil cylinder, weapon mount, and inner shield.

Above: *Jagdpanzer* IV made in late April-early May 1944, on display in Saumur, France. To save more weight, the lower sides of inner armor for the ball mantlet were shortened. The front basic armor of the vehicle was still 60 mm thick. The makers of the armor plate removed the mount for the machine gun near the driver.

Right: The curved sight channel with its cover.

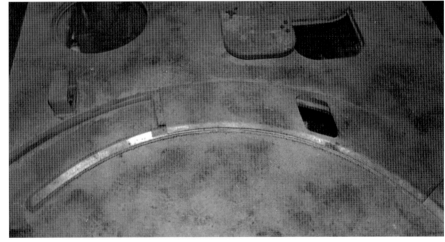

Details of the cannon with the muzzle brake removed.

A look from the left at the cannon and elevation apparatus.

A look at the driver's place from the aiming gunner's seat. The big handwheel was for elevation, the small one for traversing the primary weapon. The round opening in the left sidewall is the filler for the front fuel tank.

The mount for the Sfl ZF 1a targeting scope.

An ammunition storage rack on the right side of the vehicle. Next to it is the SSG 76 transmission. The fuel tank lies under the ammunition rack.

The hull brackets for the (movable) ammunition rack to the commander's left. The vertical toothed rod holds the shear telescope.

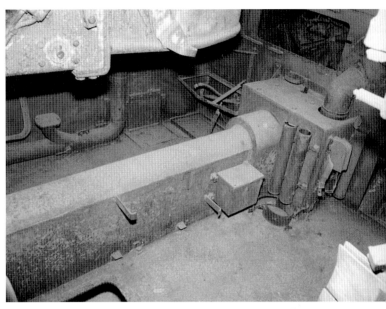

A look from the aiming gunner's position toward the firewall of the engine compartment. Plus the covering of the cardan shaft.

A view of the attachment for the 7.5 cm Pak 39 on the roof of the body.

A look at the loading gunner's ventilation seat at the firewall, the heating ducts, and the connection to the heat exit vent in the engine compartment.

January 1944

For all *Jagdpanzer* IV, a round opening was planned in the roof to hold the close-range defensive weapon. For production reasons, though, this weapon was not always available in sufficient quantities. Thus, a round armor-plate cover with four screws closed the opening on most *Jagdpanzer* IV.

February 1944

To make the front end of the vehicle lighter, the spare track links, which had formerly been stowed on the upper front plate, were moved to the rear of the body. The two spare road wheels that were originally carried on the upper end of the hull received new brackets on the left engine compartment cover.

March 1944

The left machine-gun port in the upper bow plate was removed, because unlimited use was not possible. Since the opening was already burned into the steel, the assembly firm welded a conical plate (60 mm thick) in this round opening until bodies without this hole could be delivered.

Some *Jagdpanzer* made in March and April 1944 were fitted with an experimental mount for the "all-round-fire machine gun" in front of the loading gunner's visor.

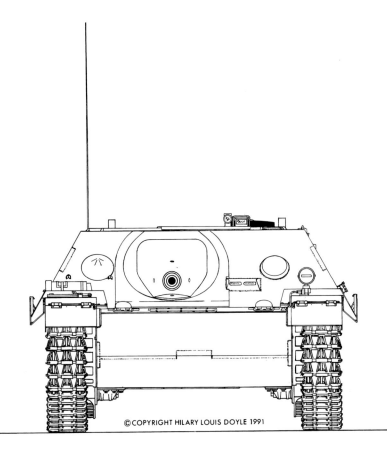

©COPYRIGHT HILARY LOUIS DOYLE 1991

Jagdpanzer IV (60 mm front armor)
March 1944 - (MG for aiming gunner removed - opening for it closed and welded - spare track links moved to the rear)

April 1944

The lower corners of the inner weapon mount attachment were removed to decrease the vehicle's nose-heaviness more.

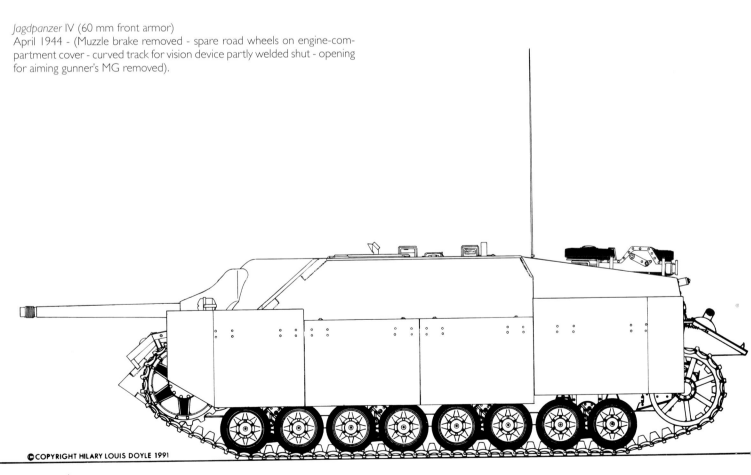

© COPYRIGHT HILARY LOUIS DOYLE 1991

Jagdpanzer IV (60 mm front armor)
April 1944 - (Muzzle brake removed - spare road wheels on engine-compartment cover - curved track for vision device partly welded shut - opening for aiming gunner's MG removed).

© COPYRIGHT HILARY LOUIS DOYLE 1991

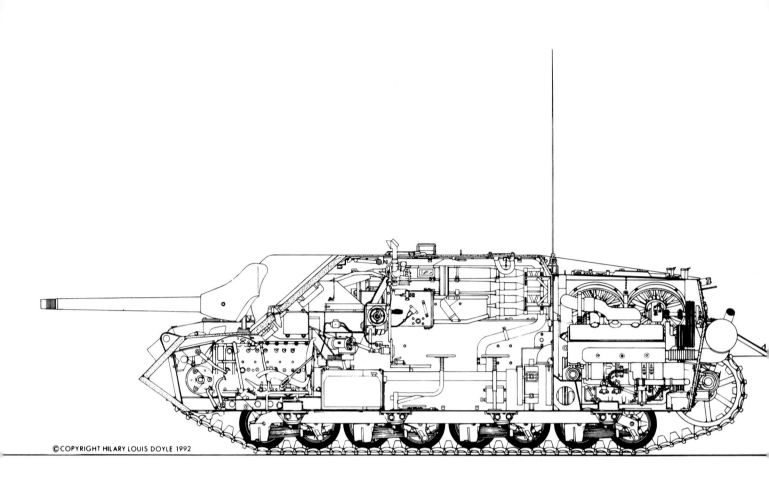

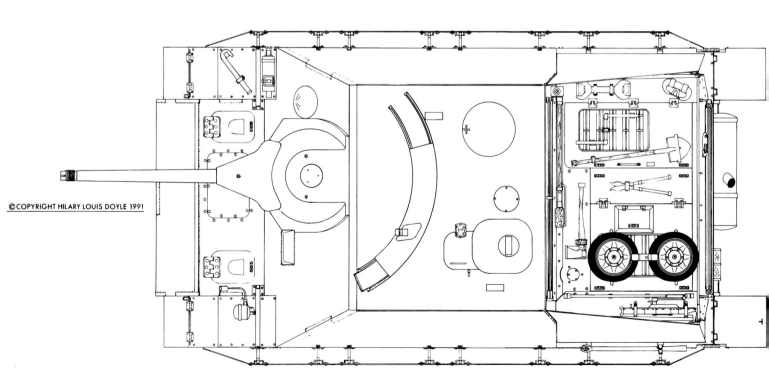

122

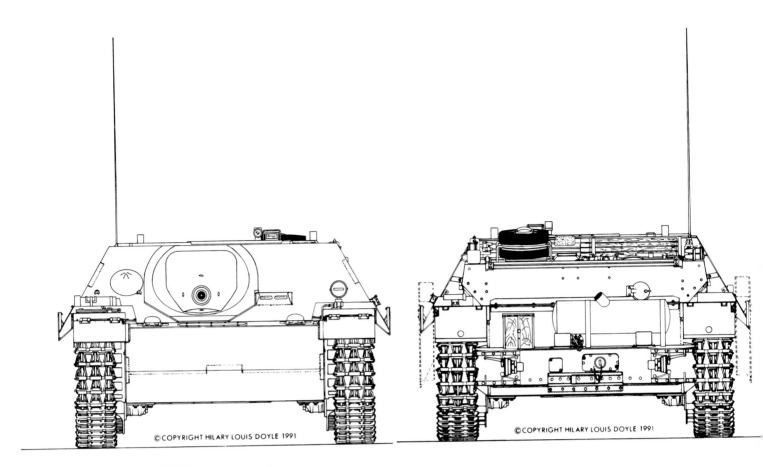

Jagdpanzer IV (60 mm front armor)

May 1944:

Beginning with chassis no. 320301, the armor thickness of the front hull and body plates was increased to 80 mm. The diameter of the conical cover for the machine-gun mount was enlarged, thus improving the holding ability and traversing of the weapon. By the end of May 1944 the guns of the *Jagdpanzer* IV were fitted with muzzle brakes. Because of the heavy dust when firing, which strongly influenced vision, the muzzle brakes were no longer mounted during assembly. The troops usually removed them in the field. Since, unlike being mounted in a turret, the weapon was attached to the upper bow plate, the recoil was spread over the entire vehicle.

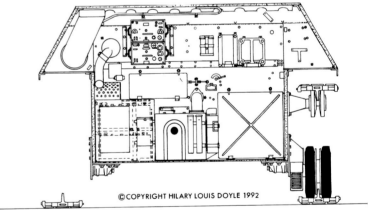

Transverse drawing, looking toward the firewall.

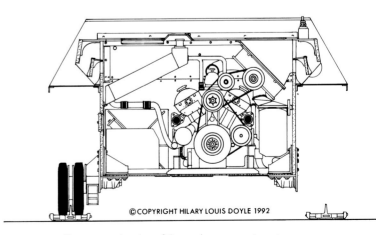

Transverse drawing of the engine compartment.

Jagdpanzer IV - Command Tank
(80 mm front armor)
May-June 1944 - (Star Antenna D - threading for muzzle brake eliminated -
curved track of vision device changed - new ball mantlet with rounded base-
plate)

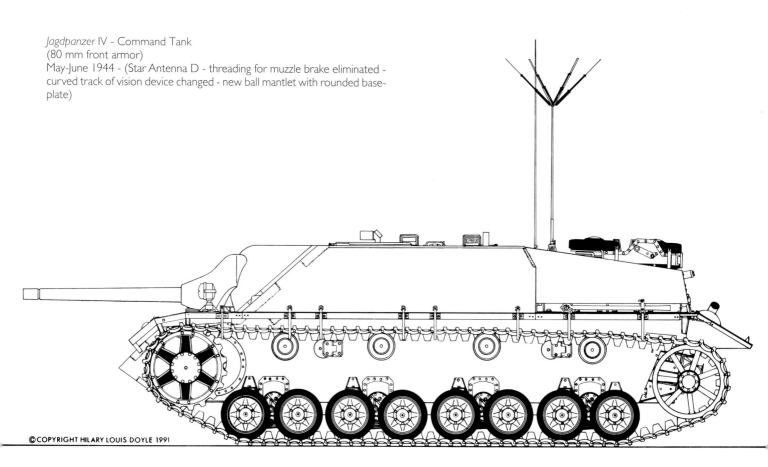

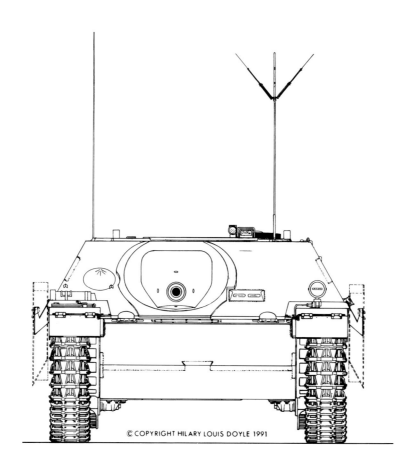

June 1944:

To make production simpler, the formerly angled sides of the armored cover over the coolant filler in the engine compartment cover were now made rectangular.

September 1944:

Newly designed, vertically mounted "flame-killer" exhaust pipes were attached to reduce flame formation.

To spare a roller bearing and shorten production time, the number of jack rollers per side was reduced from four to three.

Troop complaints about water getting in when it rained resulted in the attachment of small brackets that were welded onto the front, sides, and rear of the body. They allowed a tent canvas to be attached to cover the entire roof.

Rumor had it that the anti-magnetic "Zimmerit" protective covering led to fires when hit by shells, even if the armor was not pierced. Thus, the assembly firms no longer applied "Zimmerit."

Leaders' vehicles, as already noted, received a second antenna, readily recognizable from outside, for the additional radio equipment, mounted on the left front corner of the engine compartment cover.

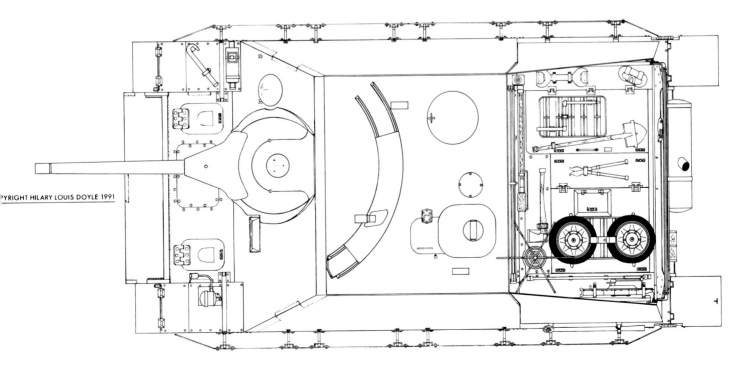

COPYRIGHT HILARY LOUIS DOYLE 1991

Jagdpanzer IV with 80 mm front armor (Chassis no. 320380). Made in June 1944 and sent to Kummersdorf for testing.

Vehicles with 80 mm front armor could be recognized from the front by the visible horizontal space between the hull and upper body.

The shape of the flange for the weapon mount was made slimmer to save weight. A cutout on the flange allowed the formation of a larger machine-gun opening that could be closed completely.

The bow of the tank as seen from above.

The roof of this *Jagdpanzer* IV, now displayed in Thun, Switzerland, shows the "close-range defense weapon" mount. The original "Zimmerit" protective coating was removed from this vehicle.

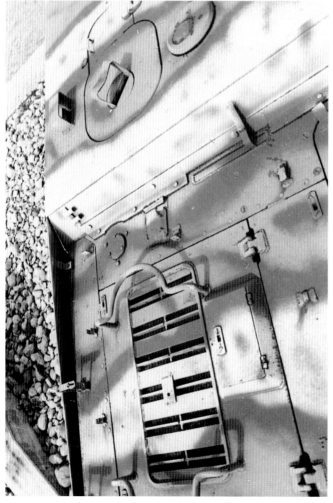

Details of the engine compartment cover. The attachment for the second antenna is already foreseen, in case the vehicle is used as a command tank.

Production

In January 1944, Vomag had finished the first thirty *Jagdpanzer* IV, which the Army Weapons Office accepted. At this time Vomag was still building Panzer IV tanks, whose production continued until May 1944. The production change from tank to *Jagdpanzer* IV followed smoothly. As the following overview of production and acceptance shows, the production increased steadily to over 100 units in April, and reached its height of 140 vehicles in July 1944. With the introduction of the long 7.5 cm Pak L/70 in August, production of the *Jagdpanzer* IV with the 7.5 cm Pak 39 L/48 ended; the last two such vehicles were built in November 1944.

Jagdpanzer IV Production

Month 1944	Ordered	Accepted	Notes
January	50	30	
February	60	45	production difficulties
March	90	75	production difficulties
April	120	106	production difficulties
May	140	90	supplying firms bombed
June	120	120	
July	130	140	
August	80	92	
September	60	58	bomb attack
October	54	46	bomb attack
November	0	2	

Service

As of March 1944, *Jandpanzer* IV were delivered to the *Panzerjäger* units of the armored and armored grenadier divisions. Every *Panzerjäger* company received ten or fourteen *Jagdpanzer* IV, in accordance with the organization plan stated in K.St.N. 1149. In most cases, the *Panzerjäger* unit of an armored division had two companies, each with ten *Jagdpanzer* IV, and one more vehicle for the unit commander. The *Panzerjäger* unit of a *Panzergrenadier* division had two companies, each with 14 *Jagdpanzer* IV, plus three more *Jagdpanzer* for the unit staff.

Exceptions also proved the rule here. The *Panzerjäger* Instructional Unit 130 of the Armored Instruction Division was the first unit to receive the *Jagdpazner* IV. It was originally planned to have only one company with 14 *Jagdpanzer* IV, and one company with 14 *Jagdtiger*. Because of production delays with the *Jagdtiger* these were not delivered. A reorganization of the *Panzerjäger* Instructional Unit 130 finally assigned nine *Jagdpanzer* IV to each of three companies, and another four *Jagdpanzer* IV for the unit staff.

The second exception was the "Hermann Göring" Armored Paratroop Division, which received *Jagdpanzer* IV for the third unit of its armored regiment in April 1944. A second assignment of *Jagdpanzer* IV to this division included ten vehicles for each of the three companies. In addition, one *Jagdpanzer* IV was assigned to the commander of the "Hermann Göring" Paratroop *Panzerjäger* Unit.

The rest of the *Jagdpanzer* IV, which were not assigned directly to the combat units, went to the Army Weapons Office for testing, or to the various schools for training.

During the Allied landing in Normandy on June 6, 1944, the units in the west had only 62 *Jagdpanzer* IV available (31 with the Panzer Instructional Division, 21 with the 2ⁿᵈ Panzer Division, and ten with the 12ᵗʰ SS Panzer Division). The remaining 21 *Jagdpanzer* IV promised to the 12ᵗʰ SS Division "Hitler Youth" were not released by the Army Equipment Office until June 22, 1944. The following units in the west were likewise given *Jagdpanzer* IV: the 17ᵗʰ SS Panzer Division "Götz von Berlichingen," the 116ᵗʰ Panzer Division, the 9ᵗʰ Panzer Division, the 11ᵗʰ Panzer Division, and the 9ᵗʰ SS Panzer Division "Frundsberg" (in the order listed). At the beginning of the Ardennes offensive on December 16, 1944, there were still 92 *Jagdpanzer* IV in service with the Panzer and *Panzergrenadier* divisions in the west.

Three German divisions that fought against the Allies in Italy received a total of 83 *Jagdpanzer* IV by April 1944.

Comparison
Jagdpanzer IV against Enemy Tanks

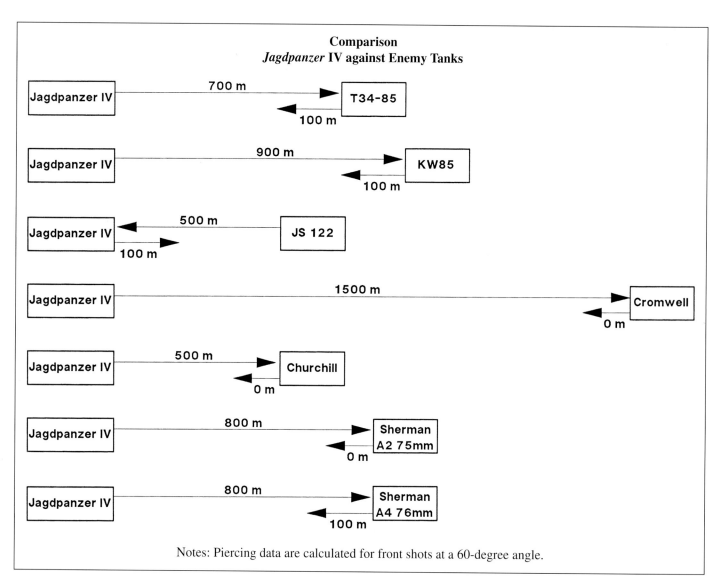

Notes: Piercing data are calculated for front shots at a 60-degree angle.

Twenty-one of them went to the "Hermann Göring" Armored Paratroop Division, and 31 to the *Panzerjäger* units of the 3rd *Panzergrenadier* Division and the 15th *Panzergrenadier* Division.

The 21 *Jagdpanzer* IV of the III. Unit of the "Hermann Göring" Armored Regiment were the first to see action. All other units equipped with *Jagdpanzer* IV were assigned to the eastern front.

The availability of the *Jagdpanzer* IV is shown by the following strength report of December 30, 1944:

Front	Division	Orig. (ready for action)	New Arrivals
Eastern			
H.G. South	3rd Pz. Div.	12 (5)	21 in August
	6th Pz. Div.	10 (5)	10 in July, 11 in August
	8th Pz. Div.	12 (8)	21 in September
	20th Pz. Div.	19 (16)	21 in June, 21 in September
	23rd Pz. Div.	14 (8)	21 in October
	3rd SS Pz. Div.	17 (16)	21 in September
	5th SS Pz. Div.	13 (13)	21 in July
H.Gr.A	19th Pz. Div.	21 (21)	10 in July, 11 in August
	25th Pz. Div.	21 (20)	28 in August
H.Gr. Center	4th Pz. Div.	12 (2)	21 in July
H.Gr. North	Pz. Div. H.G.	18 (13)	31 in July
	5th Pz. Div.	25 (11)	21 in June, 21 in July
	12th Pz. Div.	15 (8)	21 in June
	Total	209 (146)	311
Western			
H.Gr.B	Pz.Lehr.Div.	0 (0)	31 in March
	2nd Pz. Div.	1 (1)	21 in April
	9th Pz. Div.	6 (1)	21 in July, 10 in September
	116th Pz. Div.	9 (4)	21 in July
	9th SS Pz. Div.	1 (0)	21 in July
	12th SS Pz. Div.	0 (0)	10 in April, 11 in June
	3rd Pz.Gren.Div.	7 (0)	31 in April
	15th Pz.Gren/Div.	19 (12)	31 in May, 6 in August
H.Gr. G	11th Pz. Div.	14 (10)	21 in August, 20 in November
	17th SS Pz. Gr.Div.	2 (0)	31 in June
	Total:	59 (28)	286

These figures show clearly that the chances of *Jagdpanzer* surviving in the east were very much higher than in the west. There were several reasons, such as terrain, enemy tank strength, and tank defense, plus the absolute air superiority of the Allies in the west. Most decisive was the much greater combat experience of the German units in the east.

Experience Reports

The "Nachrichtenblatt der Panzertruppen" of November 1944 gives the following look at the action of *Jagdpanzer* IV on the eastern front:

Nachrichtenblatt der Panzertruppe
November 1944
Experiences of a
Panzerjäger Unit in the East with *Jagdpanzer* IV

1. The Jagdpanzer IV has proved itself fully against fire from Russian 7.62 cm Pak, *Panzerbüchsen*, and grenade launchers. Of the unit's 21 *Jagdpanzer* IV, no one was lost to enemy activity despite several hits from these weapons.

The task of securing our own units against enemy tanks and supporting the grenadiers in infantry combat could be fulfilled in every case. In offensive combat, the holding of large sectors without support from other troop units was also temporarily successful.

Subordination to smaller units than the regiment easily leads to divided action, and thus to unnecessary losses. It is thus necessary that the leader of a *Jagdpanzer* IV unit apply the closed action of his unit with clear suggestions, and does not let leadership be taken from his hands.

2. The order was given from higher up to take *Jagdpanzer* that were not ready for service into combat as well, and to build in no longer mobile *Jagdpanzer* as fixed antitank guns.

The carrying out of this order had to lead to the loss of these *Jagdpanzer*.

For this reason, the leader of a *Jagdpanzer* IV unit must make clear with all energy that the action of an immobile *Jagdpanzer* is pointless, since he cannot turn the chassis without a running motor and—robbed of its mobility—is easy prey to the enemy, or must be blown up.

It is also to be made clear that most technical damage can be repaired in hours or a few days, so that the affected *Jagdpanzer* is then fully ready for action and available to the troops, while it is lost in carrying out such orders.

A responsible unit leader must therefore try by all means to have vehicles not ready for service free for repairs.

3. The use of *Jagdpanzer* IV in unobservable terrain without observation by grenadiers often leads to the loss of the vehicle to the action of enemy antitank forces.

The constant subordination of a grenadier unit to a *Panzerjäger* (assault gun) unit has worked splendidly. The strengthened grenadier company assigned to the unit has, ever since the first combat, fit in fully through their means of escorting and watching over the *Jagdpanzer* against antitank forces. It was also in a position to make independent attacks and counterthrusts with limited goals, under the fire support of the *Jagdpanzer*, and to put down shock troop undertakings by sighted enemy heavy weapons that could not be fired upon by the weapons of the *Jagdpanzer*.

4. The fighting of infantry targets with explosive shells, that are desired only for their effect on the grenadiers' morale, is not to be approved, what with the short supply of ammunition.

The bow machine gun is splendidly effective at all target ranges when the commander maintains a strict fire control.

The use of explosive shells should be extended to firing on sighted heavy weapons, against closed units, and to self-defense. For self-defense, reserves of all ammunition must be defined as untouchable. It is suggested that the shells in the racks to the left of the commander of a *Jagdpanzer* IV be defined for this purpose, and one belt each of M.G. ammunition (1:5), ammunition for pistols and machine pistols, and five egg hand grenades be reserved.

5. The action of *Jagdpanzer* units without sufficient repair service and means of towing leads to unnecessary losses of vehicles.

The independent action of such units or single vehicles is thus to be refused, as long as technical service by repair personnel of similar units is not included.

6. In unscouted situations, such as resulted during the combat around Baranowicze and in the following transfer movements, damaged tanks of other units were pushed off roads, ordered to be used for supplying and, since they were too slow, were reached and blown up by the enemy.

In this case it did not turn out to be purposeful to apply repair services in the vicinity of railroad lines and to tow damaged vehicles there (made mobile by wheeled vehicles) under the direction of an energetic and alert officer, so that loading would be possible under the threat of enemy action. In this way the unit was able, by reaching damaged vehicles, to rescue 17 *Jagdpanzer* IV, three Panzer III, three Panzer IV, and one Hungarian tank from the enemy by loading them onto railroad cars on lines well outside their sector.

7. Through appropriate training, and under the direction of an energetic officer, it has been possible in every case, even under heavy enemy fire, to remove equipment from a *Jagdpanzer* IV prepared for being blown up and have it rescued by the crew, sometimes despite being surrounded by the enemy. Thus, the equipment from two *Jagdpanzer* IV was taken through the enemy at night.

Removing the equipment, including the radios and transformers, plus preparation for blowing up and carrying it out, must be taught as part of training.

8. Medical treatment of badly wounded men in mobile combat has proved to be difficult, especially at night. The unit had no ambulances or medical armored vehicles. Thus, the following is suggested:

In night combat, a trench about one meter deep is dug under the commander's vehicle, supplied with blankets and tent canvas, and prevented from showing light by blankets. In this excavation the medical officer can treat the wounded in full light while protected from enemy action.

Standpoint of the inspector-General of the Armored Troops

The report shows that the *Jagdpanzer* IV can, with the effects of its weapons and its armor, fulfill its main tasks: fighting against tanks and supporting the infantry.

Provided, though, that it is used according to its unique nature.

Correct use of the mobility of the *Jagdpanzer* and the great firepower of closed units, as well as considering the necessity of technical service, bring decisive success, and also guarantee a long maintenance of readiness for action of the valuable device.

Jagdpanzer IV New Assignments

Month	Transported*	Total	Division	Unit	Front
1944					
March	17.March	31	Pz.Lehr Div.	Pz.Jg.Lehr Abt.130	Western
	4.Apr.	14	2. Pz.Div.	Pz.Jg.Abt.38	Western
April	12. Apr.	7	2.Pz.Div.	Pz.Jg.Abt.38	Western
	26. Apr.	10	12.SS Pz.Div.	Pz.Jg.Abt.12	Western
	22. Jun.	11	12.SS Pz.Div.	Pz.Jg.Abt.12	Western
	25. Apr.	21	Fs.Pz.Div.H.G.	Ill./Pz.Rgt.H.G	South
	29. Apr.	31	3.Pz.Gren.Div.	Pz.Jg.Abt.3	South
	30. Mai	31	15.Pz.Gren.Div.	Pz.Jg.Abt.33	South
Mai	6. June	21	4.Pz.Div.	Pz.Jg.Abt.49	Eastern
	11. June	21	5.Pz.Div.	Pz.Jg.Abt.53	Eastern
	30. June	31	17.SS Pz.Gren.	Pz.Jg.Abt.1 7	Western
June	19. June	21	20.Pz.Div.	Pz.Jg.Abt.92	Eastern
	24. June	21	12.Pz.Div.	Pz.Jg.Abt.2	Eastern
	10. July	21	116.Pz.Div.	Pz.Jg.Abt.228	Western
	20. July	21	9.Pz.Div.	Pz.Jg.Abt.50	Western
	1. Aug.	21	1 1.Pz. Div.	Pz.Jg.Abt.61	Western
	18. July	10	19.Pz.Div.	Pz.J g.Abt. 19	Eastern
July	8. July	21	5.SS Pz.Div.	Pz.Jg.Abt.5	Eastern
	13. July	10	6.Pz.Div.	Pz.Jg.Abt.41	Eastern
	29. July	21	9.SS Pz.Div.	Pz.Jg.Abt.9	Western
	3. Aug.	21	3.Pz.Div.	Pz.Jg.Abt.543	Eastern
	29. July	21	5.Pz.Div.	Nachschub	Eastern
	26. July	31	Fs.Pz.Div.H.G.	Fs.Pz.Jg.Abt.H.G.	Eastern
	1. Aug.	11	19.Pz.Div.	Pz.Jg.Abt.19	Eastern
August	9. Aug.	11	6.Pz.Div.	Pz.Jg.Abt.41	Eastern
	22. Aug.	21	10.SS Pz.Div.	Pz.Jg.Abt.10	Western
	13. Aug.	28	25.Pz.Div.	Pz.Jg.Abt.87	Eastern
	5. Sept.	21	ISS Pz.Div.	Pz.Jg.Abt.3	Eastern
	19. Aug.	6	15.Pz.Gren.Div	Nachschub	Western
September	11. Sept.	21	8.Pz.Div.	Pz.Jg.Abt.43	Eastern
	8. Sept.	10	9.Pz.Div.	Nachschub	Eastern
	24. Sept.	21	20.Pz.Div.	Nachschub	Eastern
October	13. Oct.	21	23.Pz.Div.	Pz.Jg.Abt.128	Eastern
	31. Oct.	31	4.SS Pz.Gren.	Pz.Jg.Abt.4	Eastern
November	11. Nov.	14	1 1.Pz.Div.	Nachschub	Western
	20. Nov.	6	1 1.Pz.Div.	Nachschub	Western

* Sent out by the Army Equipment Office

Panzer IV/70 V*

Development

The original discussions by the Army Weapons Office in September 1942 about a *Jagdpanzer* on Panzer IV chassis set the requirement for armament with the 7.5 cm L/70. The 7.5 cm *Panzerjägerkanone* 42 L/70 ranks among the outstanding antitank weapons of World War II. Developed as a tank gun (KwK) with a muzzle brake for the Panther tank, it provided (through modifications to the barrel recoil) a version without a muzzle brake for *Jagdpanzer*. It was a semi-automatic weapon with electric ignition. It fired antitank and explosive shells (*Patronenmunition*).

* According to D 97/1 Device 559 - *Panzerwagen* 604/10

Vomag built the prototype of the Panzer IV/70 (V) with a modified 7.5 cm KwK 42 L/70 on the basis of the *Jagdpanzer* IV. The vehicle was introduced in April 1944, and still had the threading for the muzzle brake, which could be dispensed with because of the greater space in the fighting compartment. The vehicle still had the 60 mm front armor. The machine-gun mount at the left front was welded shut.

Measures, Weights, and Performance Data

I. Gun barrel

Caliber in mm	75
Barrel length in caliber	70
Barrel length in mm	5225
Charge space length in mm	668
Gas pressure	ca. 3200 kg/sq/cm/
Rifling:	
Number of riflings	32
Depth of riflings	0.9 mm
Width of riflings	3.86 + 1 mm
Field width	3.5-1 mm
Rifling type	uniform
Riflings in degree	6 deg. 30 min.

II. Performance

Ammunition:	
7.5 cm PzGrPatr. 39/42	
Shot weight	6.8 kg
Muzzle velocity	925 m/sec
7.5 cm PzGrPatr. 40/02	
Shot weight	4.66 kg
Muzzle velocity	1070 m/sec
Maximum range	10000 m

Despite the superiority of this weapon, it was not originally considered for the *Jagdpanzer* IV. This vehicle was put into production in January 1944 with the 7.5 cm Pak 39 L/48. At the Führer's conference of January 25-27, 1944, the installation of the 7.5 cm Pak L./70 was discussed again. This long cannon was to be built into the "*Panzerjäger* Vomag" as soon as the technical problems were solved and sufficient production was assured.

At the beginning of April 1944 the Army Weapons Office introduced a rebuilt 7.5 cm KwK 42 L/70 that was built into a *Jagdpanzer* IV (chassis no. 320162). On April 6, 1944, Hitler was shown several photos of this vehicle. He was sure that this *Jagdpanzer* with the 7.5 cm Pak L/70 was the last word in tank technology. On April 20, 1944, Hitler was shown the vehicle. He arranged for it to be added to the *Jagdpanzer* program with a final monthly production of 800 vehicles. The long-term program of the Army Weapons Office as of May 4, 1944, called for 2020 *Jagdpanzer* IV of both types in the period from April 1944 to April 1945.

On July 18, 1944, Hitler ordered that this vehicle be designated Panzer IV long (V). The added "V" indicated development by the Vomag firm, also the only producer of this vehicle. To differentiate this *Jagdpanzer* from the Panzer IV with the long 7.5 cm KwK L/43 or L/48, the troops preferred the designation Panzer IV/70 (V). The Army Weapons Office accepted this designation officially only in November 1944.

The *Jagdpanzer* (chassis no. 320162), still designated Panzer IV lang (V), was tested thoroughly at Kummersdorf. An "all-round-fire machine gun" was mounted on the roof of the body. A folding barrel brace for the cannon was attached to the upper bow plate.

A Panzer IV/70 (V), made by Vmag in August 1944. The "Zimmerit" protective coating was applied after this picture was taken. The ends of the side aprons were bent inward to avoid being turned off. The cannon still had the threading for the muzzle brake.

This Panzer IV/70 (V) (chassis no. 320756), finished and sent out in September 1944 as "Panzer IV/L (V)," was delivered to Panzer Brigade 107 or 108. The vehicle had rubber-tired road wheels and four steel jack rollers.

Development of Designations for the Panzer IV/70 (V)

Sturmgeschütz auf Fahrgestell **Panzer IV mit der 7.5 cm L/70**
Führer's conference   10/2/1942
Panzerjäger - **Vomag mit 7.5 cm L/70**
Führer's conference   4/6 and 4/19/1944
Pz.Jäg. Vomag mit 7.5 cm Pak L/70 and
Sturmgeschütz **n.A. mit 7.5 cm Stu.K.L/70**
GenStdH/General der Artillerie Kriegstagebuch   6/7/1944
Stu.Gesch.n.A. mit 7.5 cm Pak L/70 auf Fgst.Pz.Kpf.Wg.IV
"Overview of the Army's Armament State" Chef H Rüst u. BdE/
Stab Rüst III   6/15- 7/15/1944
le.Pz.Jg.IV Vomag mit L/70
Gen.Insp.d.Pz.Tr.Akten   6/27/1944
Sturmgeschütz **mit der langen 7.5 cm L/70**
Führer's conference   7/6/1944
All *Sturmgeschütze* **mit 7.5 cm L/70** were, at the Führer's order,
designated
"Panzer IV lang." The *Sturmgeschütz* **auf Pz.Jg.Vomag**
Fahrgestell was designated
"Panzer IV lang (V)."   7/18/1944
Panzer IV lang (V) m. 7.5 cm Pak 42 L/70
"Overview of the Army's Armament State" Chef H Rüst u. BdE/
Stab Rüst III   8/15 - 10/15/1944
le.Pz.Jg. (Vomag) mit 7.5 cm Pak 42
Wa A Abn.   8/31/1944
Panzer IV lang (V)
Chef H Rüst u. BdE, Wa.Abn.   9/6/1944
Le.Pz.Jg. Vomag mit 7.5 cm Pak L/70 auf Fgst. Pz.
IV (lang Vomag)
Chef GenStdH/Org.Abt./Gen.Insp.d.Pz.Tr.   9/8/1944
Called by the troops
Panzer IV lang; called officially
Panzer IV lang Ausf,V (Vomag)
Chef GenStdH/Org.Abs./Gen.Insp.d.Pz.Tr.   9/11/1944
Jagdpanzer **IV**
Gen.Insp.d.Pz.Tr.Akten   10/19 - 12/4/1944
Panzer IV lang (V)
Gen.Insp.d.Pz.Tr.Akten   12/10 1944-4/6/1945
Panzer IV/70 (V),
Panzerwagen 604/10 (V) (m.7.5 cm Pak42 L/70)
"Overview of the Army's Armament State" Chef H Rüst u. BdE/
Stab Rüst III   11/15/1944-3/15/1945
Panzer IV lang (V): Verskraft   1945

Specific Features

In order to revise the *Jagdpanzer* IV for the new gun, a number of changes were necessary. The inner and outer gun covers had to be changed to save weight, but the armor protection could not be decreased. An added external barrel brace would hinder adjusting the means of sighting in rough and heavy terrain. The long overhang of the barrel proved to be a problem. Instead of simple body ventilators, a barrel exhaust blower was installed to blow out the powder smoke immediately after firing. The storage for the longer and larger-diameter ammunition of the 7.5 cm Pak 42 L/70 was also changed. To save weight, the ammunition rack originally mounted in the right bow area was eliminated. Only 55 rounds could be carried.

The *Jagdpanzer* IV with the 7.5 cm Pak 39 L/48 was already nose-heavy. This disadvantage was made even worse by the installation of the long cannon. WaPrüf 6 suggested on May 16, 1944, that the whole suspension be lengthened 100 mm to the front, which would also have changed the center of gravity for the better. But this change required a new design for the bow. The front road wheel mounts were already quite close to the drive wheel. Any further extension forward would inevitably have led to a collision of the two running-gear components, which would have caused the destruction of the first road wheel by the drive wheel.

On August 10, 1944, a further suggestion called for limiting the front armor, and the use of rubber-saving steel road wheels on both front trucks. On August 11, 1944, Hitler ordered the reduction of the front armor to 60 mm, an order that was never carried out. During the whole production the front armor remained 80 mm thick. To solve the problem, the two front wheel trucks were simply fitted with steel road wheels, and lighter tracks were introduced. Both changes were made in September 1944. One of the two spare road wheels carried on the rear body was also changed to a steel wheel.

Further changes made in August and September 1944 corresponded to those made on the *Jagdpanzer* IV.

Above: This Panzer IV/70 (V), made early in October 1944, was sent to the 1st SS Armored Division. The picture shows the vehicle in action during the Ardennes offensive. The Sfl. SF 1a scope for the aiming gunner and the shear scope for the commander are easy to see. The number of road wheels on each side was reduced to three; the front trucks had steel road wheels. The camouflage paint was applied by the manufacturer.

Right: Details of the steel road wheels. A rubber ring was pressed in between the outer steel rim and the wheel rim. As of September 1944 the two upper central bolts to hold the wheel truck were eliminated.

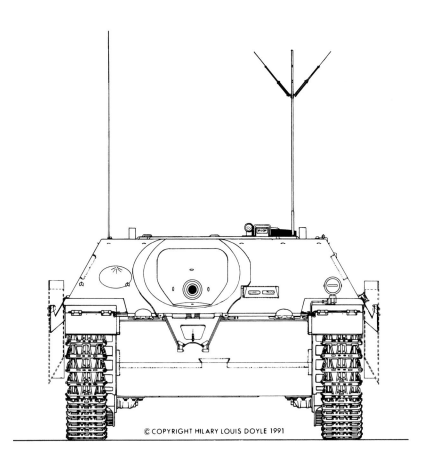

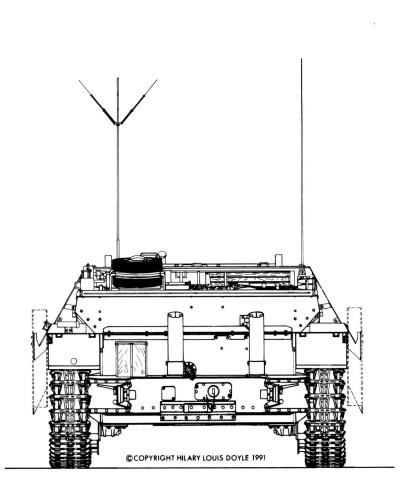

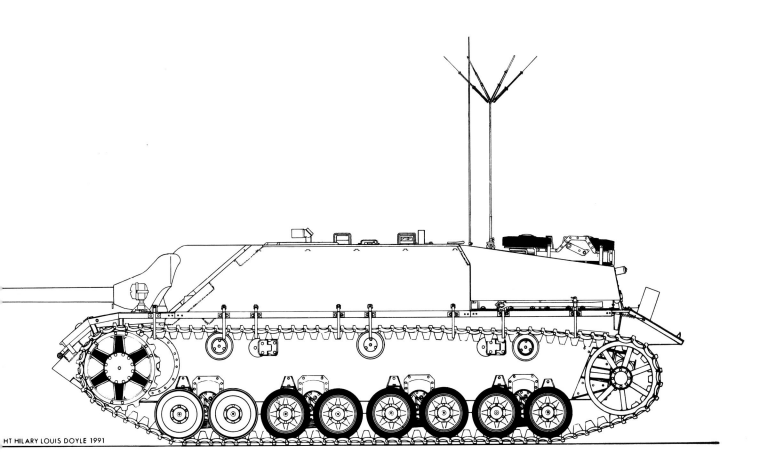

Panzer IV/70 (V) - Command Tank
September 1944 - (Front wheel track with steel road wheels - new shortened arc for visor - new ball mantlet - cannon bracket with springs - fuel filler
with square lid - flame extinguisher - new light tracks)

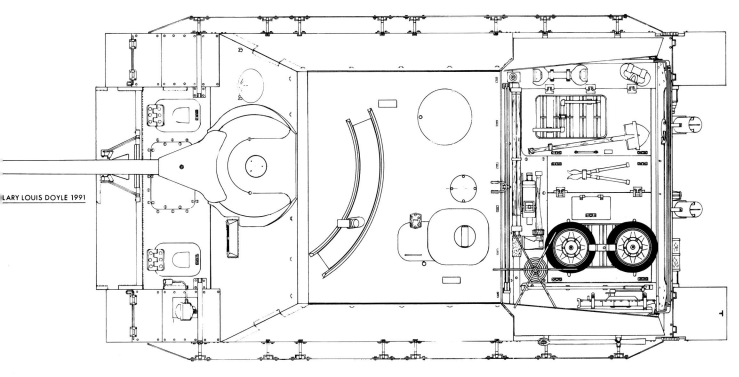

The diameter of the opening for the bow machine gun (closed here) was made so much smaller that the cutout on the flange of the mount could be eliminated.

A look through the opening for the bow machine gun on the gun mount.

A look at the machine-gun mount from inside.

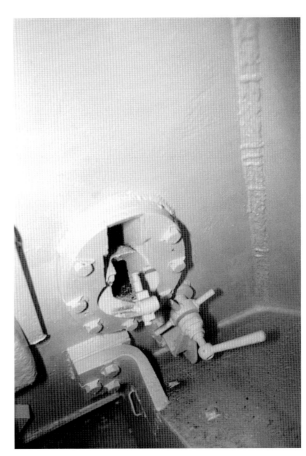

Below: The engine compartment cover of the Panzer IV/70 (V) shows the changed mount for the second antenna. Thus, the vehicle could be used as a command tank. The bracket for the barrel-wiper rods was moved, as was the flame extinguisher bracket from the right front track cover. This vehicle is at Aberdeen, MD, USA.

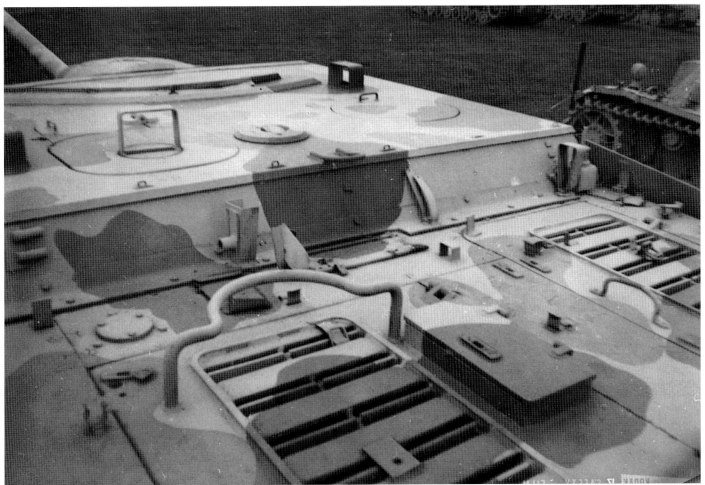

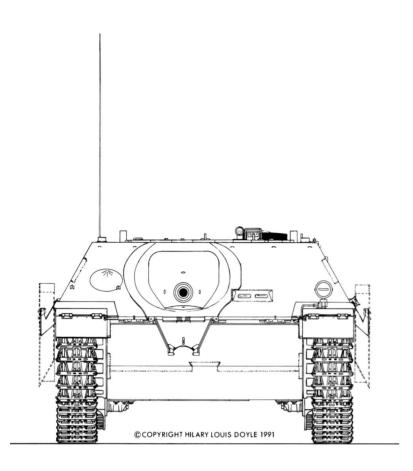

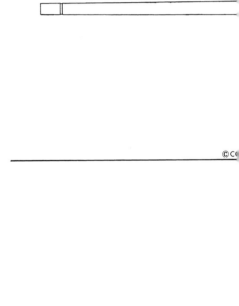

© COPYRIGHT HILARY LOUIS DOYLE 1991

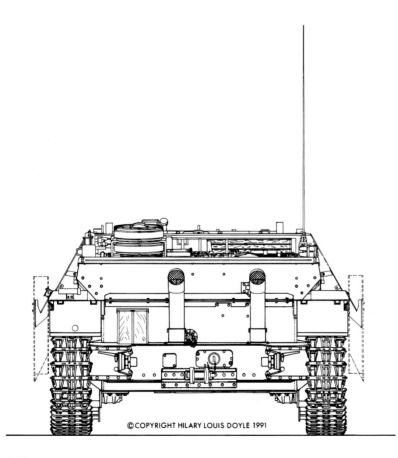

© COPYRIGHT HILARY LOUIS DOYLE 1991

Panzer IV/70 (V)
March 1945 - (Attachments for the two-ton crane - flame extinguisher cover - rubber-tired spare road wheel - close combat weapon - Sfl ZF 1a scope - towing coupling - ventilators on the servicing flaps on the upper bow plate eliminated - new driver's scope (periscopes in various directions) - improved cannon mounting)

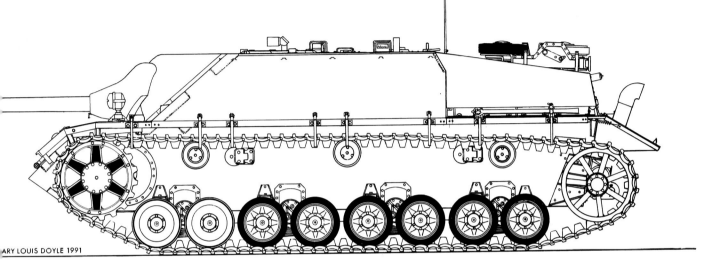

ARY LOUIS DOYLE 1991

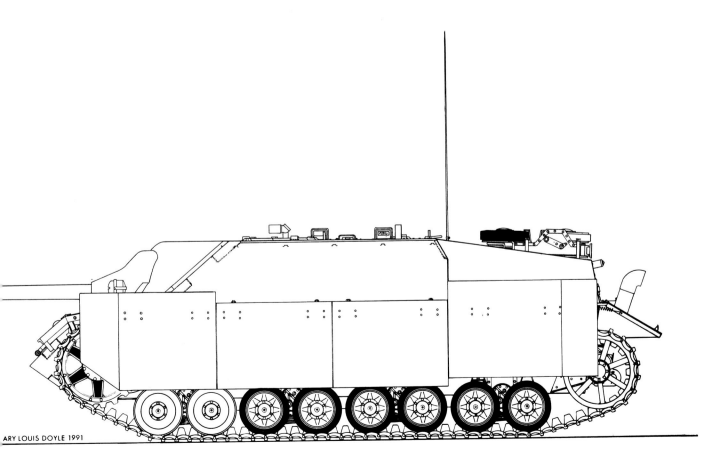

ARY LOUIS DOYLE 1991

Further visible differences appeared only in November, when the central towing coupling on the rear plate was welded on, and "mushrooms" for attaching the two-ton crane to the roof were added. The roof also added three rods to serve as brackets for a range finder. The ventilators on the upper front plate were eliminated, since it had been found on November 4, 1944, that the air scoops needed to cool the brakes of the Panzer IV were useless on the Panzer IV/70. The brakes received air from an air shaft, through which smoke and heat from the brakes were sucked backward to the engine compartment and discharged through the cooling air ventilator. This change had already been made in the first *Jagdpanzer* IV.

While all *Jagdpanzer* IV and most Panzer IV/70 (V) used the leading wheels of the Panzer IV Type F tank, which were welded together of tubes, some Panzer IV/70 (V) were fitted with still-available cast leading wheels of the Panzer IV Type H in February and March 1945.

Panzer IV/70 (V)

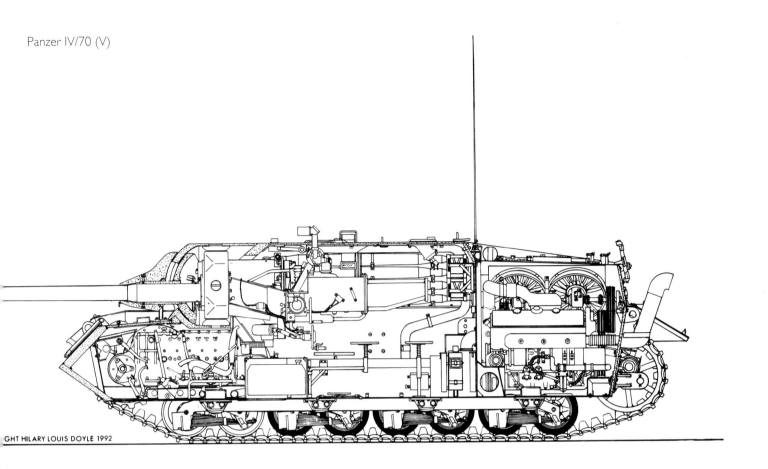

GHT HILARY LOUIS DOYLE 1992

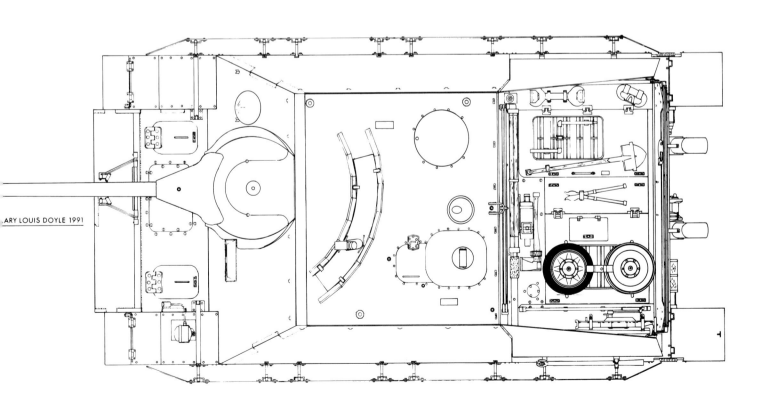

ARY LOUIS DOYLE 1991

147

Production

Vomag produced the first 57 Panzer IV.70 (V) in August 1944, and the Army Weapons Office accepted them in the same month. Production continued in September with 41 units, and expanded to 104 in October 1944. 178 were built in November, and 180 in December 1944. The greatest number (185 vehicles) was attained in January 1945. Problems with spare parts supplying and energy failures reduced the production to 135 Panzer IV/70 (V) in February and 50 in March.

The bombing raids on Plauen on March 19, 21, and 23, 1945, ended production once and for all, after a total of 930 Panzer IV/ 70 (V) had been built.

A Panzer IV/70 (V) with the newer, wider driver's window, captured by British troops, is still on display in Britain.

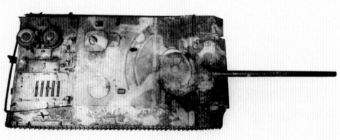

148

Rear view of a Panzer IV/70 (V) (chassis no. 329667), made early in March 1945. The vehicle shows the vertical towing eyes applied in December 1944, and the lighter tracks introduced in September 1944. It is on display in Ottawa, Canada.

Below: A captured Command Tank IV/70 (V) with equipment for a company chief. On the left edge of the engine compartment cover was the increased-range antenna for the FuG 8 radio. The cast leading wheels, and the fact that only the forward road wheels have steel rims, are unusual. The brake-cooling intakes on the upper bow plate were replaced by handholds. The paint was applied by the manufacturer.

Service

The newly established Panzer brigades were the first units to receive the *Jagdpanzer* IV/70 (V). Every armored unit of these brigades was to include three companies of Panthers, while the fourth—the *Panzerjäger* company—had eleven Panzer IV/70 (V). The 105th and 106th Panzer Brigades received eleven Panzer IV/70 (V) each in August 1944. In September 1944, eleven were sent to each of five Panzer brigades (107th, 108th, 109th, 110th, and the Führer Grenadier Brigade). Panzer Brigades 105 to 108 saw service in the west, while Brigades 109 and 110 were assigned to the east. In September, another ten Panzer IV/70 (V) were sent as replacements to the 116th Panzer Division in the west, and in October another ten went to the 24th Panzer Division on the eastern front.

This company, still ready for action, surrendered to the Canadians in May 1945. The close-up of the roof shows the viewer heads for the aiming gunner's targeting scope and the commander's shear scope. Three rods—one forward and two side by side behind the commander's scope—were introduced in 1945 to serve as a bracket for the commander's planned EM 9 range finder. The roof of the vehicle behind it clearly shows the welded-on "mushroom" brackets for the two-ton crane. This unit's second vehicle served as a command tank. The side aprons were omitted for lack of material.

Inside view of a Panzer IV/70
(V) on display at Aberdeen,
MD, USA:

1) Driver's area
2) Loader's machine gun,
   ammunition box open
3) Loader's hatch
4) Radio rack
5) Commander's area "close-
   combat defense weapon"
   in roof extension
6) Commander's area,
   vertical attachment for the
   Sfl 4 Z shear scope.

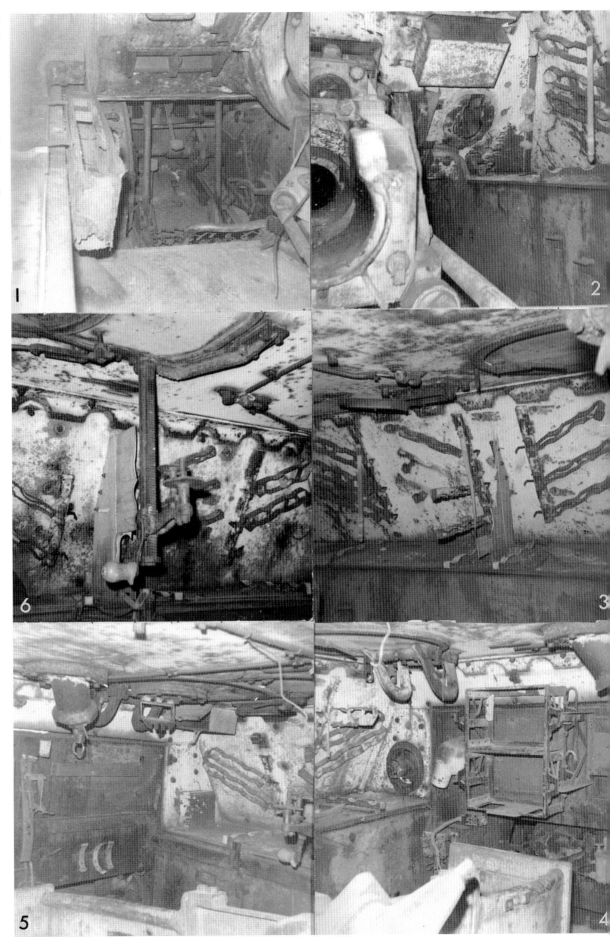

The majority of the supplying in the months of October to December 1944 were sent off to refresh those *Panzerjäger* units that had been intended to take part in the Ardennes offensive in December 1944, and the subsequent "Nordwind" operation in January 1945. Every *Panzerjäger* company within the armored divisions was to receive ten Panzer IV/70 (V), while every company in the armored grenadier divisions and the heavy Army *Panzerjäger* units hoped for 14 of them.

In order of delivery as arranged by the Army Equipment Office, the following units received Panzer IV/70 (V):

Number	Transported	Unit
6	10/6/1944	s.H.Pz.Jg.Abt. 560
21	10/20/1944	1. SS Pz.Div.
21	10/21/1944	12. SS Pz.Div.
7	10/25/1944	s.H.Pz.Jg.Abt. 560
21	11/9/1944	9. SS Pz.Div.
21	1/13/1944	Pz.Lehr Div.
15	11/14/1944	s.H.Pz.Jg.Abt. 560
20	11/18/1944	2. SS Pz.Div.
28	11/25/1944	s.H.Pz.Jg.Abt. 655
4	11/28/1944	9th Pz.Div.
3	12/2/1944	s.H.Pz.Jg.Abt. 560
9	12/4/1944	s.H.Pz.Jg.Abt. 519
17	12/4/1944	3. Pz.Gren.Div.
16	12/7/1944	s.H.Pz.Jg.Abt. 559
3	12/7/1944	s.H.Pz.Jg.Abt. 655
3	12/9/1944	10. SS Pz.Div.
10	12/14/1944	9. Pz.Div.
5	12/14/1944	116. Pz.Div.
11	12/14/1944	Pz. Lehr Div.
2	12/14/1944	s.H.Pz.Jg.Abt. 559
7	12/15/1944	10. SS Pz.Div.
5	12/21/1944	15. Pz.Gren.Div.
22	12/26/1944	25. Pz.Gren.Div.
17	12.27.1944	21. Pz.Div.

At the beginning of the Ardennes offensive. 210 Panzer IV/70 (V) were with the units in the west, while another 90 reached the troops before the offensive ended.

In the latter half of December the main interest was turned again to the hard-fighting eastern front. The *Panzerjäger* units of the 7th, 13th, and 17th Panzer Divisions each received 21 Panzer IV/70 (V) as replacements for their *Panzerjäger* units, while the 24th Panzer Division received 19 vehicles.

The situation on all fronts got progressively worse, and the official organization plans went unnoticed. Efforts to fill the ever-increasing gaps on all fronts remained piecemeal. Armored vehicles were sent out to troop units that were available. There was no time to train the crews of the new vehicles. They had to see service as soon as they had left the transport train. This took place in the following units:

Heavy Panzer Unit 563: 31 Panzer IV/70 (V) in January 1945; II.Unit/Panzer Regiment 9: 26 Panzer in January 1945; Infantry Division Döberitz: 10 Panzer in February 1945; Panzer Unit 303 (Silesia): 10 Panzer in February 1945; *Panzerjäger* Unit 510: 10 Panzer in February 1945; Panzer Unit Jüterbog: 10 Panzer in February 1945; SS *Panzergrenadier* Division "Nordland": 10 Panzer in March 1945.

Most of the remaining Panzer IV/70 (V) were sent to the eastern front as replacements, and reached their units between January and March 1945 as follows:

Division	Number	Reached Unit
Pz.Gren.Div.GD	21	January 1945
10. Pz.Gren.Div.	10	January 1945
21. Pz.Div.	6	February 7, 1045
7. Pz.Div.	17	February 8, 1945
17. Pz.Div.	28	February 8, 1945
2. SS Pz.Div.	8	February 17, 1945
1. SS Pz.Div.	11	February 18, 1945
9. SS Pz.Div.	12	February 18, 1945
12. SS Pz.Div.	21	February 18, 1945
25. Pz.Div.	10	February 21, 1945
10. Pz.Gren.Div.	10	February 2, 1945
20. Pz.Gren.Div.	20	February 26, 1945
25th Pz.Div.	10	March 2, 1945
8. Pz.Div.	10	March 5, 1945
20. Pz.Div.	10	March 1945
13. Pz.Div.	8	March 21, 1945

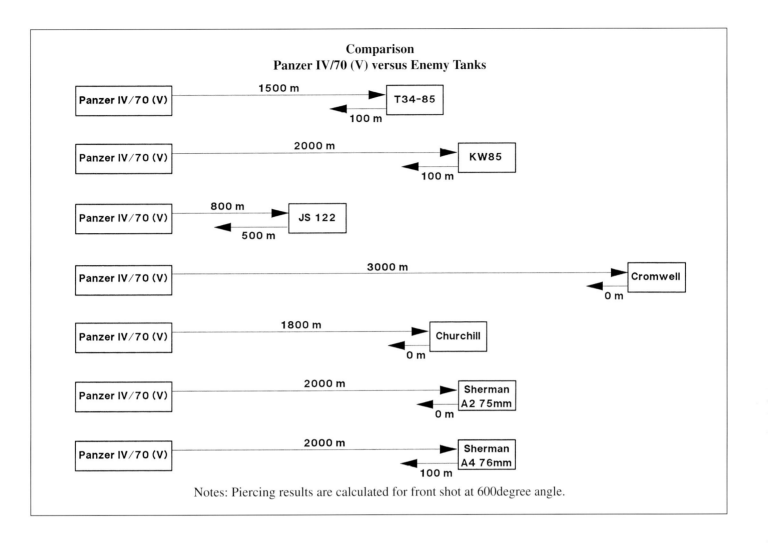

Comparison
Panzer IV/70 (V) versus Enemy Tanks

Notes: Piercing results are calculated for front shot at 600degree angle.

In a desperate attempt to stop the front in the west, the last 59 Panzer IV/70 (V) were sent as replacements to various units at the front. Seventeen reached the main battle lines on 16 March, 21 on 3 April, 20 on 4 April, and the last on 8 April.

One of the few documents that involve the action of Panzer IV/70 (V) has survived as the diary of a platoon leader of the 6th Company, Panzer Regiment 9:

"At the end of January 1945 the company received 14 Panzer IV/70 (V) at the Neuhammer troop training facility in the vicinity of Sagan, Silesia. The vehicles were checked briefly, camouflaged for winter use, test-driven, and the cannons were fired. The new *Jagdpanzer* were the best armored vehicles in which I ever sat or fought during the war. Especially in antitank defense, which had absolute priority in the last months of the war.

We saw action in the last days of January; after many contacts with the enemy, the unit was assigned to hold bridgehead on the Elbe near Stettin. The platoon was assigned to an infantry unit to strengthen the defense and carry on reconnaissance.

On March 16, 1945, the Russians began to prepare for an attack with heavy artillery fire and constant waves of 50 to 60 bombers. It was the heaviest fire concentration that I ever experienced on the eastern front. We immediately retracted our shear telescopes and machine guns and closed all the openings. The shell fragments were flying around, but could only scratch our armor.

About 9:00 AM we learned that the Russians had gathered numerous tanks before our front infantry lines. From radio contact with our unit and the regiment, we finally learned from an infantry messenger that the rest of our company and unit had attacked already.

Moving forward, we were delayed by the terrain, which was torn up by bombs and shells. Just at 1:00 PM the artillery fire suddenly ended. A deadly quiet prevailed around us. Then balls of light rose into the sky from the foxholes and MG emplacements—the enemy is attacking.

The first T 34/85 and SU 85 tanks rolled into sight of our *Jagdpanzer*, which waited in concealed firing positions. One saw the first flashes of light from hits on the leading T 34s, which began to smoke shortly thereafter. Then five to eight more enemy tanks appeared behind them, which also began to burn at once. The same thing happened to all the attacking enemy tanks. Every shot from our cannons was now a hit. It was not in vain that we detailed our most experienced and oldest corporals and sergeants as aiming gunners; they scarcely missed their targets. After some 30 minutes of firefight, a strong formation of T-34 tried to get around us to the side and attack from the right flank. We had used almost all of our ammunition when other guns opened fire on the attacking Russian vehicles from beside and behind us. The rest of the unit had caught up to us, and supported our bitter defensive battle against the overpowering Russian tank formations.

Our *Jagdpanzer* took numerous hits during this action, which lasted well into the afternoon. But we stayed in action until the last surviving Russian tank had withdrawn. During the retreat of the enemy vehicles we took one more heavy hit on the intermediate gears, which immobilized our vehicle. We waited for our friends from the tank recovery unit, who towed us to the other bank of the Oder."

In a discussion with the Chief of Staff on March 14, 1945, about questions of development that concerned further developments and necessity programs resulting from the existing shortage of raw materials, *Generaloberst* Heinz Guderian explained:

"As the Inspector-General of the Panzer Troops sees it, in the present situation, or at the time when the actual 'need program' began, 75 'Panther' tanks, even if they embody the most modern and refined state of design, are no more valuable than 150 'Panzer IV/70 (V),' with all this vehicle's weaknesses. Instead of 'Panthers,' 'Jagdpanthers' are needed. Numbers depend on steel supplies."

After the Panther tank had turned out to be the best German war vehicle built in World War II, it is a great compliment for the Panzer IV/70 to be evaluated so positively by Guderian.*

* This was not always true. In a speech note in 1943, Guderian raised the following question: g) Another testing of the necessity of designing a light assault gun with the 7.5 cm Cannon L/70. In that case, drop this plan and substitute the light assault gun with 7.5 Cannon L/48 and armored troop carriers. The value of the 7.5 cm assault gun L/70 is as yet untested.

Panzer IV/70 A*

Development

The Panzer IV/70 (A) is one of the German *Wehrmacht's* least-known armored vehicles. It was pushed into the background by the *Jagdpanther*, the *Jagdtiger*, and even its sister vehicle, the Panzer IV/70 (V). In postwar publications it usually appears as a "temporary solution." The few available pictures and the only surviving example, on display in the French Army's tank museum,

show an overly rushed design without regard for shot deflecting form or the whole body of the vehicle.

On June 24, 1944, technicians at the Army Weapons Office in Hillersleben finished tests of the survival capability of the Panzer IV tank versus the Russian T 34/85 and JS 12. The Panzer IV showed itself to be very inferior by all criteria. On June 26, 1944, the Altmärkische Kettenwerk GmbH received a contract to install the long 7.5 cm KwK L/70 in a Panzer IV chassis as soon as possible. Similar tests had already been finished in September 1943, and showed five reasons why the 7.5 cm Cannon L/70 was not

* According to D 97/1 Device 558 - *Panzerwagen* 604/9

Alkett finished the prototype of the Panzer IV/70 (A) in a short time in June/July 1944. The normal body of a *Jagdpanzer* IV was mounted on an unchanged Panzer IV chassis. To give the cannon the necessary elevation area without changing the position of the fuel tanks, a vertical collar was welded in between the chassis and body. The conical cover for the bow machine gun and the welded-shut driver's visor show that a normal *Jagdpanzer* IV body was used.

suitable for installation in the Panzer IV turret. The decision was made that—in view of all the disadvantages—the body of the Panzer IV/70 (V) should be modified to fit on the Panzer IV chassis.

A second comparison study of the shot ranges of the cannons with their penetrating power, this time including American and British tanks, confirmed the fact that in this area the Panzer IV was inferior to all new enemy tanks. A report of July 5, 1944, noted that the rebuilding of the Panzer IV into an "Assault Gun with 7.5 cm KwK 42 L/70" would give superiority over all enemy tanks, with the exception of the "Josef Stalin 122." The concepts of assault gun and tank destroyer were thus interchangeable. While the artillery preferred the assault gun concept, the armored forces used the *Jagdpanzer* concept.

Development of Designations for the Panzer IV/70 (A)

Umbewaffneter Pz/Kpfw.IV mit L/70
Gen.Insp.d.Pz.Tr.Akten   6/27/1944
All *Sturmgeschütze* mit 7.5 cm L/70 were, on orders from the Führer, designated
"Panzer IV lang." The *Sturmgeschütz* auf **Pz.Kpfw.IV Fahrgestell** was designated
"Panzer IV lang (A)"   7/18/1944
Panzer IV lang (A) m. 7.5 cm Pak 42 L/70
"Overview of the Army's Armament State" Chef H Rüst u.BdE/
Stab Rüst III   8/15 to 10/15/1944
le.Pz.Jg. (Ni-Werk) mit 7.5 cm Pak 42
Wa A Abn.   8/31/1944

Panzer IV lang (A)
Chef H Rüst u. BdE, Wa.Abn.   9/6/1944
Panzer mit 7.5 cm Pak L/70 auf Fgst.Panzer IV mit Aufbau
Panzerjäger **Vomag** als
"Panzer IV/L (A)" (lang, Alkett)
Chef GenStdH/Org.Abt./Gen.Insp.d.Pz.Tr.   9/8/1944
Called by the troops: **Panzer IV lang**
Called officially:
Panzer IV lang Ausf. A (Alkett)
Chef GenStdH/Org.Abt./Gen.Insp.d.Pz.Tr.   9/11/1944
Panzer IV lang
Gen.Insp.d.Pz.Tr.Akten   10/19 to 11/28/1944
Panzerwagen **604/9 (7.5 cm Pak 42 (L/70)**
K.St.N. 1177   11/1/1944
Panzer IV/70 (A),
Panzerwagen **604/9 (A) (m. 7.5 cm Pak 42 L/70)**
"Overview of the Army's Armament State" Chef H Rüst u. BdE/
Stab Rüst III   11/15/1944 to 3/15/1945
Panzer IV lang (A)
Gen.Insp.d.Pz.Tr.Akten   12/4/1944 to 4/6/1945
Panzer IV lang (A) Verskraft   1945

Specific Features

Unlike the prototype, which had a low body with vertical lower plates, the latter were tilted at 20 degrees in production. The upper part of the body was like the Vomag type, whereby the weapon mounting, the weapon shield, and angle and thickness of the upper front body plate, plus the layout of the roof remained unchanged.

This prototype was tested thoroughly at Kummersdorf. In addition, the bracket for the track cover was strengthened, and the side aprons of the *Jagdpanzer* IV were used. Light tracks replaced the heavy ones with ice grippers. Production began in September 1944.

At the Nibelungenwerk, this row of Panzer IV/70 (A) stands ready for finishing after being accepted by the Weapons Office. During this time of revision, earlier and later Panzer IV chassis with either three or four jack rollers were used. The types of towing eyes varied. The Panzer IV/70 (A) was given chassis numbers from 120301 on by the Weapons Office.

But the lower front body plate with the driver's visor, the side and rear body plates, the armor protection for the machine gun and the ammunition racks had to be changed specifically for the Panzer IV/70 (A).

The body had a height of 1020 mm, while the Vomag version measured 620 mm. This height resulted from the location of the fuel tank in Panzer IV under the turret turning rail. If the normal Panzer IV/70 (V) body had been set on the Panzer IV chassis, the 7.5 cm *Panzerjägerkanone* 42 could have been mounted with only a limited elevation range, as the fuel tank otherwise would have blocked the cannon deflector. There were other reasons against raising the body:

- The inner gun mount interfered with access to and maintenance of the SSG 76 drive.
- Th 7.5 cm *Panzerjägerkanone* 42 mounted in the low body of the Panzer IV/70 (V) hit its muzzle on the ground in rough and hilly country if it was not lashed down firmly. Damage to the aiming machinery resulted.

So as not to interrupt production, the original Panzer IV chassis was not changed, and thus the greater height of the vehicle was accepted in the bargain.

As with all other German armored vehicles that had been produced over a long period of time, there were also numerous improvements to the Panzer IV/70 (A). Here only the significant changes and the times of their introduction will be listed.

September 1944

Four rubber-saving road wheels with steel surfaces were mounted on the front trucks, so as to limit wear caused by the nose heavy vehicle. Aprons of wire mesh replaced the former side plates around 18 September. The anti-magnetic "Zimmerit" coating was eliminated.

December 1944

A new towing coupling with horizontal bolts was attached to the middle of the rear. It was attached to the lower rear plate, and allowed towing of damaged vehicles by recovery tanks via towing rods. The side hull plates were lengthened in front and back and shaped as towing eyes. The former version, which was fastened with six screws, often broke off during towing.

The number of jack rollers per side was decreased from four to three, in order to save space and manufacturing time.

Panzer IV/70 (A)
December 1944 - (two front trucks with steel road wheels - wire mesh instead of sheet metal aprons - side mesh aprons could be used in two positions so winter tracks could be put on without trouble - towing coupling)

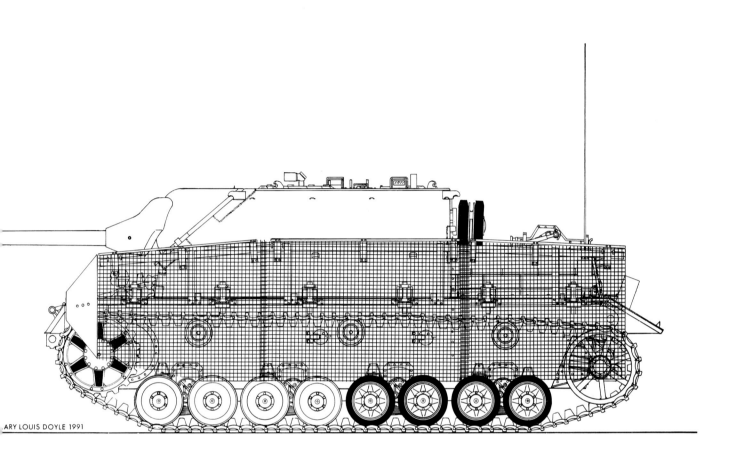

ARY LOUIS DOYLE 1991

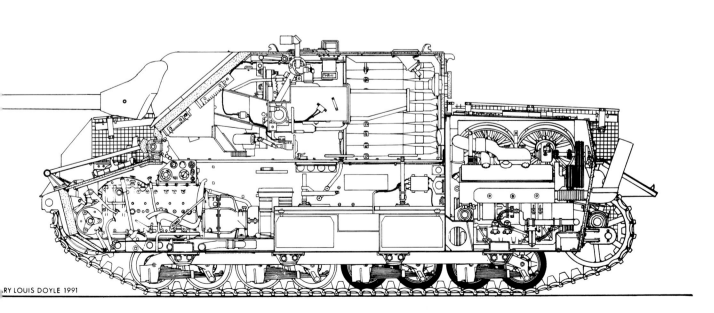

RY LOUIS DOYLE 1991

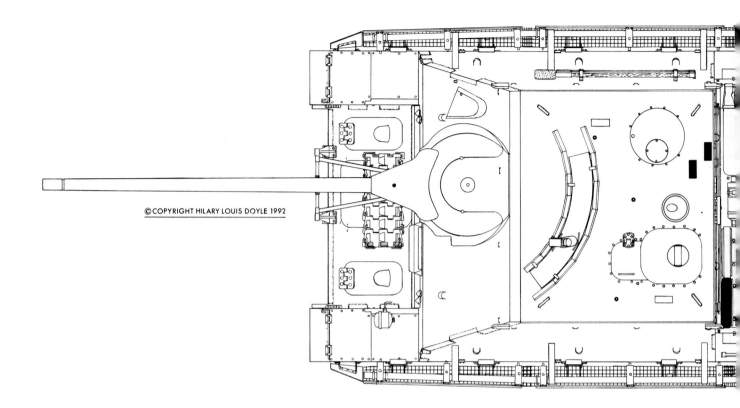

© COPYRIGHT HILARY LOUIS DOYLE 1992

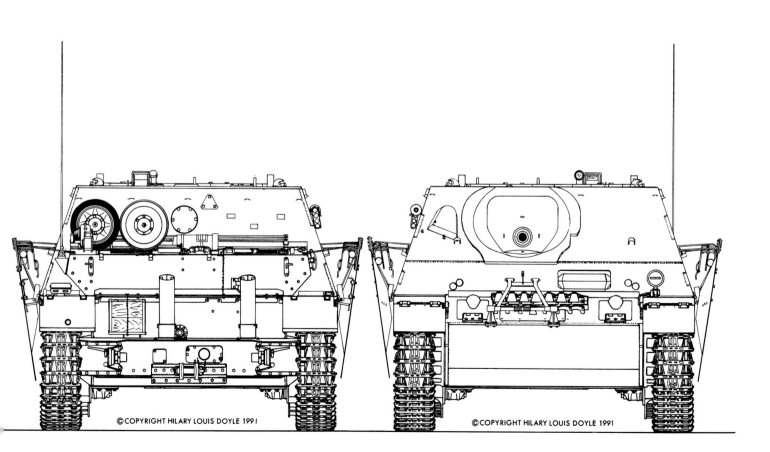

© COPYRIGHT HILARY LOUIS DOYLE 1991

© COPYRIGHT HILARY LOUIS DOYLE 1991

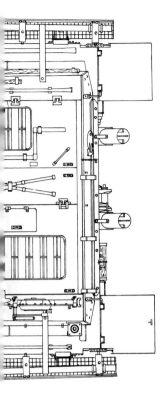

Production

The quickly assembled prototype of the Panzer IV/70 (A) designed by Alkett was shown to Adolf Hitler during the Führer's conference at the Berghof from July 6 to 8, 1944. He gave the following verdict:

As the final goal, the entire production of the *Panzer-kampfwagen* IV shall be converted to the "*Sturmgeschütz* auf Einheits-Fahrgestell III/IV" with the long 7.5 cm L/70 cannon. But since it was impossible to carry out this change at once, but on the other hand, changing to the long cannon with highest production numbers was urgent, Hitler agreed to the following transition suggestion:

-   As of August 1944, fifty of the planned 350 Panzer IV chassis are to be delivered as Panzer IV/70 (A) with transitional body, Alkett type. In order to maintain the urgently needed supplying of the newly established divisions, the delivery of these vehicles, when possible, should be moved up to the first half of August 1944.
-   As of September 1944, at least 100 of these vehicles shall be finished through further throttling of Panzer IV production.
-   The tank manufacturers were immediately ordered to convert Panzer IV production to Panzer IV/70 (V). This made it possible to deliver at least 150 more chassis equipped with the long cannon as Panzer IV/70 (V) as of October at the earliest, instead of the "transitional vehicle" Panzer IV/70 (Λ).
-   As of October 1944, fifty more modifications should be planned for every month, so that in February 1945 the whole production of 350 Panzer IV would be converted to Panzer IV/70 (V).

The Army Weapons Office at once introduced the necessary measures, but did not stick precisely to Hitler's instructions. Only the Nibelungenwerk in St. Valentin, Austria, still produced Panzer IV tanks at that time. It received the following plan for switching to production of Panzer IV/70 (A): 50 in August, 100 in September, 150 in October, 200 in November, 250 in December, and 300 in January. The contract from the Army Weapons Office stuck with the Panzer IV/70 (A) version conceived by Alkett. Switching to the Vomag version, as ordered by Hitler in October 1944, was left out at first. A second Weapons Office contract that followed immediately called for the following numbers: 50 in August, 100 in September, 150 in October, 150 in November, and 100 in December. Production would finally end in January 1945.

The change to the Uniform Chassis III/IV instead of to Panzer IV/70 (V) was commanded of the Nibelungenwerk as of November 1944.

This order was followed early in August by a third contract from the Army Weapons Office. It stated that Panzer IV production must be carried on with the increased total of 250 units per month, and a change to the Uniform Chassis III/IV would not take place.

This contract maintained the monthly production planning for 50 units in August, 100 per month from September to January, with production of Panzer IV/70 (A) continuing to February 1945.

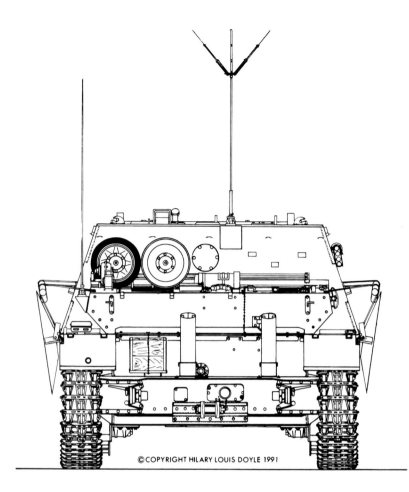

Panzer IV/70 (A) Command Car, January 1945

On January 30, 1945, the changed production plans called for 50 of the Panzer IV/70 (A) in January, 60 in February, 60 in March and April, 60 in May, and eight in June.

It can hardly be imagined why the Weapons Office, despite the ever-worsening situation, kept changing its contracts, and thus caused more confusion and delays for the producers.

The Panzer IV/70 (A), of which only 150 transitional vehicles were actually to be built, finally reached a total of 278 units:

Month	Planned	Accepted by Weapons Office	Notes
1944			
August	50	3	Difficult start, 30 finished
September	100	60	Problems at factory
October 10/16	100	43	Nibelungenwerk bombed
November	100	26	Cannon 42 delivery problems
December	80	75	
1945			
January	50	50	
February	30	20	Built only when armor was supplied
March	0	1	
Total	510	278	

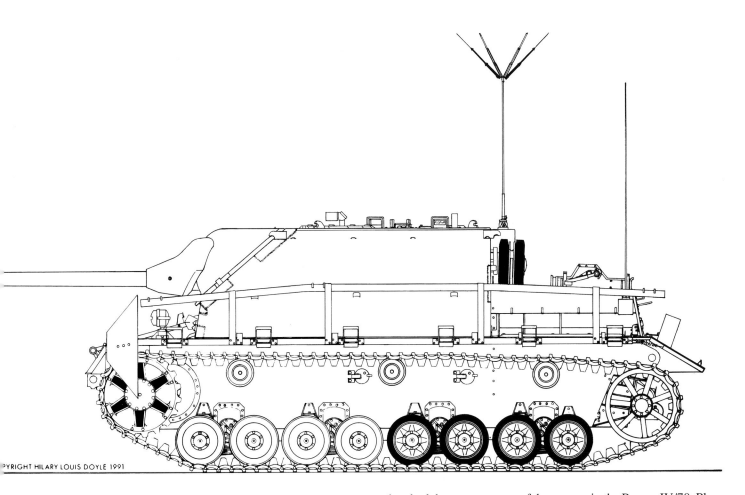

COPYRIGHT HILARY LOUIS DOYLE 1991

Action

On account of its improved firepower, the Panzer IV/70 (A) was supposed to be called on to support the Panzer IV tank. The units supplied with them could take many kinds of tactical action, as they had the greater range of the cannon in the Panzer IV/70. Plans were made in September 1944 to send 68 Panzer VI/70 (A) to units on the eastern front to strengthen them. Actually only five of these vehicles were sent out in September 1944. They went to the Panzer company of the Führer-Begleit-Brigade, which also received 27 normal Panzer IV tanks.

This Panzer IV/70 (A) was knocked out of action in France. The left front towing eye is torn off, one reason why the hull sides were lengthened and fitted with eyes. Instead of the conical cover for the machine-gun opening, series vehicles had flat discs.

Instead of side plate aprons, Panzer IV/70 (A) had aprons of wire mesh. Vehicles so equipped were used mainly on the western front. This Panzer IV/70 (A) has the standard running gear, including two forward trucks with steel-surfaced wheels, which were introduced in September 1944. The towing eyes in the side armor were used since December 1944.

Side and front views of the Panzer IV/70 (A) that French troops captured. For each type of road wheel it carried a spare.

Because of production delays, only 17 Panzer IV/70 (A) were assigned to the eastern front in September, but they were actually loaded on the train only on October 10, 1944. The remaining 51 of the original 68 *Jagdpanzer* were assigned to the eastern front in October. 34 were loaded on 8 October, and the rest (17) early in November. These Panzer IV/70 (A) were divided among the front units as follows:

| Unit | Division | Strength Reports | |
		11/1/1944	12/1/1944
8. Kp./II./Pz.Rgt.6	3. Pz.Div.	7	17
9.Kp./III./Pz.Rgt.4	13. Pz.Div.	0*)	2
6.Kp./II./Pz.Rgt.39	17.Pz.Div.	17	15
9.Kp./III./Pz.Rgt.24	24.Pz.Div.	13	7
5.Kp./Pz.Rgt.9 as			
5.Kp./Pz.Jg.Abt.87	25. Pz.Div.	17	17

According to plans of October 1944, 45 Panzer IV/70 (A) were to be given to the II. Unit of Panzer Regiment "Grossdeutschland," and another 45 to the II. Unit of Panzer Regiment 2. No longer mixed with Panzer IV tanks, 14 of them were to go to each of three companies, and three more to the unit staff. Both units were to be used as independent army units in divisions or corps—outside their own divisions—at the beginning of the Ardennes offensive.

Because of production difficulties, the II. Unit of the "Grossdeutschland" Panzer Regiment received only 38 of the assigned 45 Panzer IV/70 (A). The rest were replaced by seven Panzer IV tanks. The new vehicles reached the units at the right time for the beginning of the Ardennes offensive on December 4, 1944.

The II. Unit of Panzer Regiment 3 received only 11 Panzer IV/70 (A) on November 16, 1944. This unit received six Panzer IV tanks in November. Two companies were sent 22 "Nashorn" tank destroyers with 8.8 cm Pak 43/1 guns to fill their gaps. Assigned to Army Group G on the western front, the II. Unit of Panzer Regiment 2 was ready for action, but did not see action in the Ardennes offensive.

In December 1944 the newly established Panzer Unit 208 was equipped with 31 Panzer IV tanks and 14 Panzer IV/70 (A). The unit staff received three Panzer IV tanks: one company got all 14 Panzer IV/70 (A), while the other two each had 14 Panzer IV tanks assigned to them. In January 1945 Panzer Unit 208 was sent by rail to the eastern front, where it was subordinated to Army Group South.

The only further assignment of Panzer IV/70 (A) was to the II. Unit of Panzer Regiment 25 of the 7th Panzer Division, which belonged to Army Group Center on the eastern front; it received ten vehicles to restore its supply.

In January 1945 only two more Panzer units received the Panzer IV/70 (A). The III. Unit of Panzer Regiment 24 of the 24th Panzer Division (Army Group South) was assigned 14 of these vehicles on January 15, 1945. The I. Unit of Panzer Regiment 29 of Panzer Brigade 103, Army Group Center, was assigned 14 Panzer IV/70 (A) in January. They reached the unit by rail on February 2, 1945.

* Four reported on November 6, 1944.

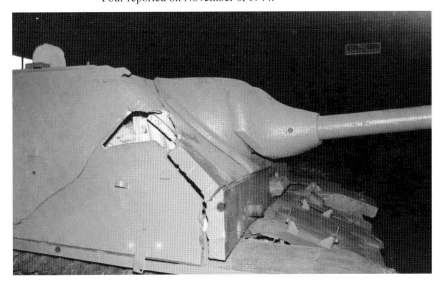

The Panzer IV/70 (A) shown here (chassis no. 120513) is now in the tank museum of the French forces at Saumur. It shows serious shot damage on the upper right side of the body, resulting from a KE hit.

This vehicle was finished at the Nibelungen Works in January 1945. The roof resembled that of the Panzer IV/70 (V). The MP 44 with curved barrel (Vorsatz P) was only installed experimentally. Rheinmetall is said to have carried out the experimental construction until November 15, 1944.

The breech of the 7.5 cm *Panzerjägerkanone* 42 L/70. The U-shaped ring at the end of the breech blew smoke and dust out of the gun barrel after firing.

Rear view of the Panzer IV/70 (A) with vertical "flame-killer" exhaust pipe.

The commander's position as seen from the loading gunner's seat. The telescope attachment is missing.

They remained the last two Panzer units to be equipped with Panzer IV/70 (A). After that, these models went only to assault gun units. The decision was made to strengthen assault-gun units with a Panzer IV/70 (A) platoon in order to be able to fire on enemy tanks effectively at greater distances. In view of the rising numbers of Sherman tanks with 76 mm and 17-pound cannons in the west, as well as the T 34/85 and JS 12 in the east, this was a sensible measure.

The aiming gunner's position, looking toward the driver. The elevation and traversing handwheels are shown clearly.

A look at the right side of the cannon shows the folding seat for the loading gunner. Brackets for an MG 42 are on the right (damaged) side of the body.

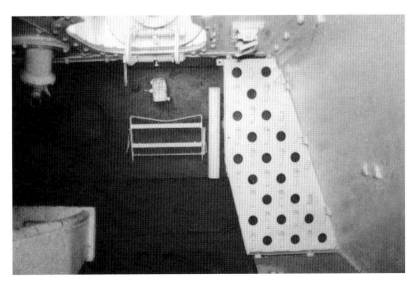

A look from the aiming gunner back to the firewall. Only the back part of the ammunition racks remained in this vehicle. Though the fighting compartment air system is missing, the "close combat defense weapon" remains.

To train the crews, the Assault Gun School in Burg, near Magdeburg, received two Panzer IV/70 (A) in mid-January 1945.

Other deliveries of Panzer IV/70 (A) to combat units were as follows:

Number	Unit	Area	Arrived
4	Stu.G.Brig. 244	West	1/26/1945
3	Stu.G.Brig. 341	West	1/26/1945
3	Stu.G.Brig. 394	West	1/26/1945
3	Stu.G.Brig. 902	West	1/26/1945
3	Stu.G.Brig. 280	West	2/13/1945
3	Sturm Artl. Brig. 905	West	2/13/1945
3	Sturm.Artl.Brig. 911	East	2/13/1945
3	Sturm Artl.Brig. 667	West	3/22/1945
3	Stu.G.Brig. 243	West	after 2/4/1945
3	Sturm Artl.Brig. 236	East	2/8/1945
3	Stu.G.Brig. 301	East	2/8/1945
31	Stu.G.Brig. "G.D."	East	2/12/1945
4	Stu.G.Brig. 300	East	3/13/1945
4	Stu.G.Brig. 311	East	3/13/1945
4	Stu.G.Brig. 210	East	3/14/1945
3	Stu.G.Brig. 190	East	3/22/1945
3	Stu.G.Brig. 276	East	3/22/1945
16	Sturm Artl.Lehrbrig.111	East	after 3/15/1945

Instead of equipping just one platoon with them, the entire "Grossdeutschland" Assault Gun Brigade received three batteries of Panzer IV/70 (A). The Assault Artillery Instructional Brigade 111 was the last unit to receive Panzer IV/70 (A), getting enough to equip four platoons.

The following combat report served as the basis for awarding the Knight's Cross of the Iron Cross to Lieutenant Hartmann of the 3rd Battery, Assault Gun Brigade 311:

"On April 18, 1945, the Russians began again with heavy artillery fire. Lt. Hartmann drove ahead to reconnoiter, and found that Russian tanks had already crossed a causeway. He put his three assault guns in alarm readiness and went back in his Fist sports car. Two more assault guns had been brought into positions during the night.

Were they already shot down by the enemy?

The three assault guns (with Hartmann in Panzer IV/70 (A)) rolled through the railroad underpass near the Obertor railroad station, and immediately came under heavy enemy artillery and grenade-launcher fire. Shortly after that, Hartmann observed through his shear telescope a number of JSU 152 assault guns. He opened fire at once and set the first one afire. More and more of these heavily armored vehicles appeared.

Hartmann fired all of his antitank shells and shot five more of these "giants" down. His other two Assault guns took part in this action. They also scored kills.

Hartmann drove back for ammunition, and noticed that the two assault guns had been placed for securing during the night; he was glad that they were still ready for action. In the meantime, his two escort vehicles had shot down five more JSU 152.

After they had loaded ammunition, all the assault guns drove to another attack. Hartmann himself shot down 13 enemy tanks and assault guns with his Panzer IV/70 (A). In all, the enemy lost 25 combat vehicles, and thus did not reach its target of Benderplatz, on the other side of the Oder. Had the Russians reached this goal, the whole island would have been lost, and the German troops would not have been able to hold Breslau."

Krupp

Suggestion for the installation of the 8.8 cm Pak L/71 in recoilless mounting
(Krupp drawing W 1792 of 11/2/1944).

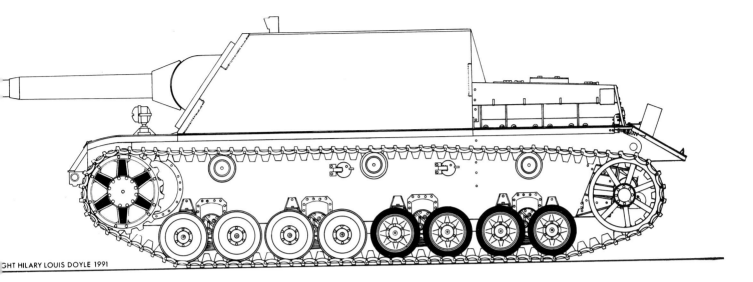

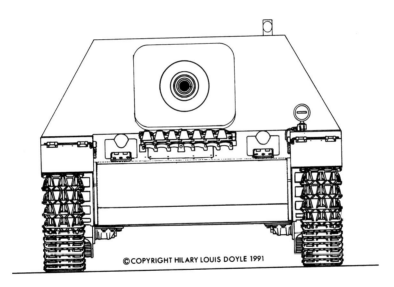

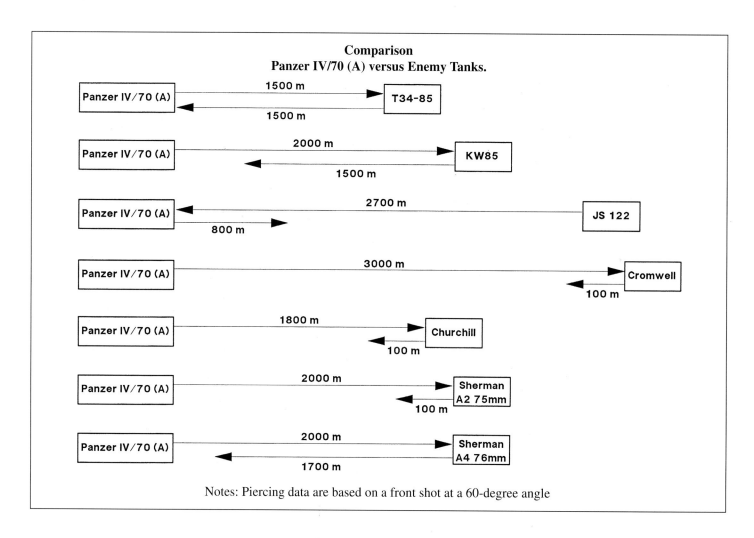

Comparison
Panzer IV/70 (A) versus Enemy Tanks.

| Panzer IV/70 (A) | 1500 m → / ← 1500 m | T34-85 |

| Panzer IV/70 (A) | 2000 m → / ← 1500 m | KW85 |

| Panzer IV/70 (A) | ← 2700 m / 800 m → | JS 122 |

| Panzer IV/70 (A) | 3000 m → / ← 100 m | Cromwell |

| Panzer IV/70 (A) | 1800 m → / ← 100 m | Churchill |

| Panzer IV/70 (A) | 2000 m → / ← 100 m | Sherman A2 75mm |

| Panzer IV/70 (A) | 2000 m → / ← 1700 m | Sherman A4 76mm |

Notes: Piercing data are based on a front shot at a 60-degree angle

This one case allows no conclusions on the actual combat value of the Panzer IV/70 (A). At that time it would have been possible at any time for the superior Allies to hold the few usable Panzer IV/70 (A) in check. If they were not shot down in direct combat, they were knocked out from the flank in enemy breakthroughs. Most of them, though, fell victim to mechanical damage and lack of fuel.

This was also the fate of the only surviving Panzer IV/70 (A), which was in the west. Knocked out by penetrating shots on the front body plates, probably from French 75 or 76 mm guns at short range, the vehicle was still driveable when captured. Today it stands in the tank museum of the French forces in Saumur.

Panzer IV (Long) E

Development

The efforts of the Army Weapons Office to unify the *Panzerkampfwagen* III and IV were practically carried out by the war's end. This "Uniform Chassis" was also supposed to be used for the light tank destroyer. In a meeting of the "Work Circle Panzer" on January 4, 1944, *Hauptdienstleiter* Saur expressed the request to search for means of realizing this state of things.

Light *Panzerjäger* and *Panzerkampfwagen* IV:

The unification was to be carried out on the basis of both the Panzer IV tank and Panzer III and IV. The weakest point of these vehicles was in both cases the intermediate gear and steering brakes.

Intermediate Gears

In the chosen combination of Panzer III/IV, a strengthening of the gears had to be made, which according to the correction suggested by Daimler-Benz gave a strengthening of 29%. In case this strengthening was still not enough, a new intermediate gear would be designed, though it could not be used to replace the old one and required other attachments.

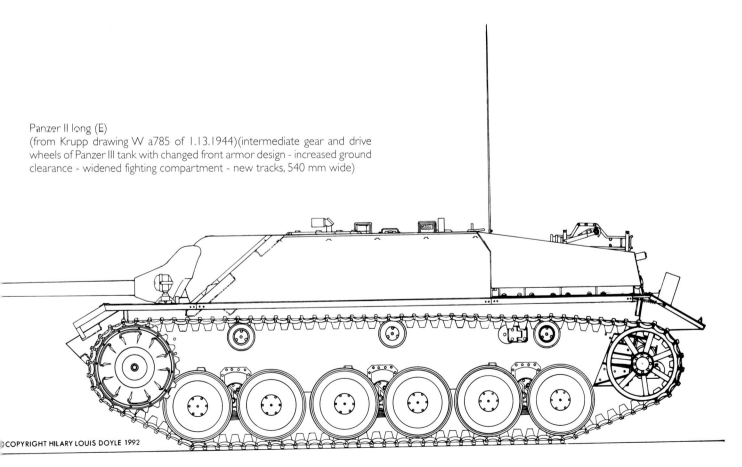

Panzer II long (E)
(from Krupp drawing W a785 of 1.13.1944)(intermediate gear and drive wheels of Panzer III tank with changed front armor design - increased ground clearance - widened fighting compartment - new tracks, 540 mm wide)

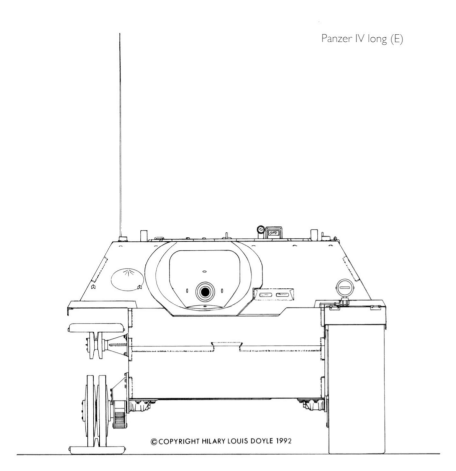

Steering Brakes

The steering brakes were at the end of their capability, as far as the braking power and activating power were concerned. The not yet concluded tests showed that the strongest steering power came from the steering brakes of Panzer IV, less from the Argus brakes, and the weakest from those of Panzer III. The Argus brakes, the essential parts of which were built like the Panther brakes, had the advantage of freedom from servicing—but the disadvantage of not definitely controllable braking power, which could lead to blocking and destroying the intermediate gear. This danger was considerably less in the hand-activation of the light *Panzerjäger* than in the Panther with its oil pressure activation.

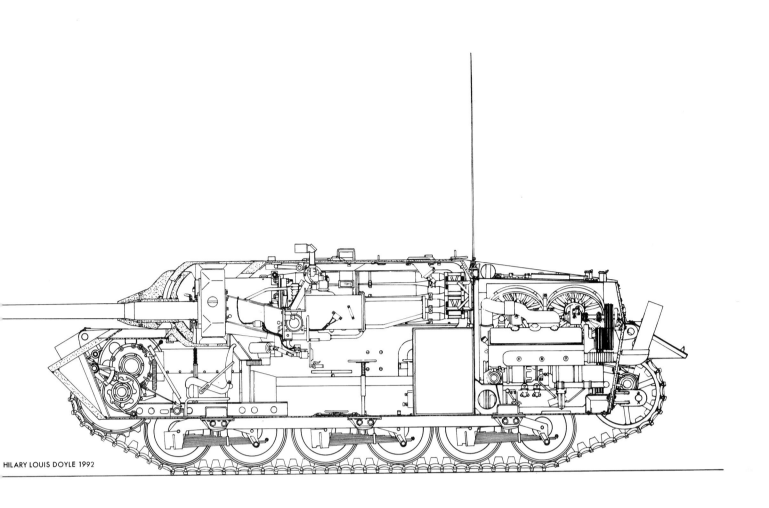

HILARY LOUIS DOYLE 1992

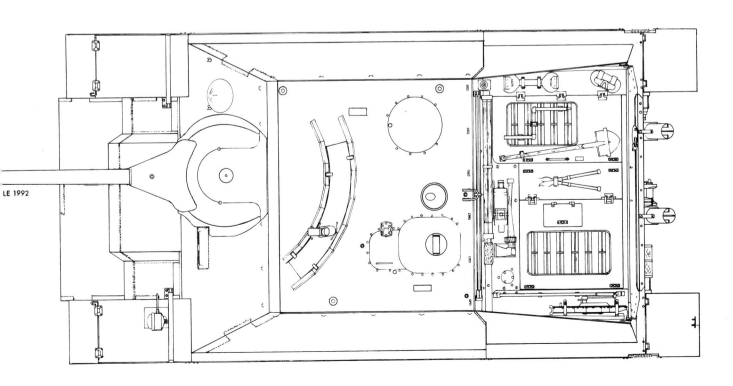

LE 1992

173

Running Gear

The running gear of the Panzer IV was, in its existing state, too weak, and could not be used any longer. Three possibilities for a thorough improvement were suggested:

- Stepped wheels with components similar to those of Panzer IV, with eight rubber-saving road wheels (700 mm diameter) suspended on four wheel trucks. The armored hull remained unchanged.
- Wheels in line, with components similar to those of Panzer IV, with six rubber-saving road wheels (660 mm diameter) suspended on three wheel trucks. The hull would have to be modified.
- Wheels in line after a suggestion by the Alkett firm, with eight rubber-saving double road wheels (470 mm diameter), suspended on four wheel trucks. The hull remained unchanged.

The first two running gear used proven components of Panzer IV, and were therefore reliable. The stepped wheels had possibilities in terms of replacement, but had the disadvantage of the not yet sufficiently known behavior of the stepped running gear in case a track came off.

The second running gear probably offered the best running characteristics. The third type, which was the easiest to manufacture—if the replacement of the roller bearings by journal bearings was possible—did not have thorough testing of its components behind it.

Cooling System

The cooling system of Panzer IV required a lot of work in construction because of the big radiators, and the two bevel drives for the ventilators. WaPrüf 6 looked into the possibility of replacing this cooling system with a simpler one. The development itself was taken on by the Dr.-Ing. h.c.F. Porsche KG. Porsche built on the cooling system for the Maus tank, and began in February 1944 to rework the Panzer III-IV chassis with the following focal points:

- Making the existing cooling system smaller;
- Improving the accessibility and simplifying the changing of the drive;
- Decreasing power losses through the cooling system.

The design (Porsche Type 260) intended to make the radiator, built by the BEHR firm, smaller. The speed of flow through the radiator was to be raised. For that a new blower, developed by the BECHT firm, was installed. The power used amounted to 6% of the engine power at 3000 rpm. The whole cooling system was mounted on the drive at the upper left, and created more room to enlarge the tank capacity from 380 to 500 liters, an increase of 31%.

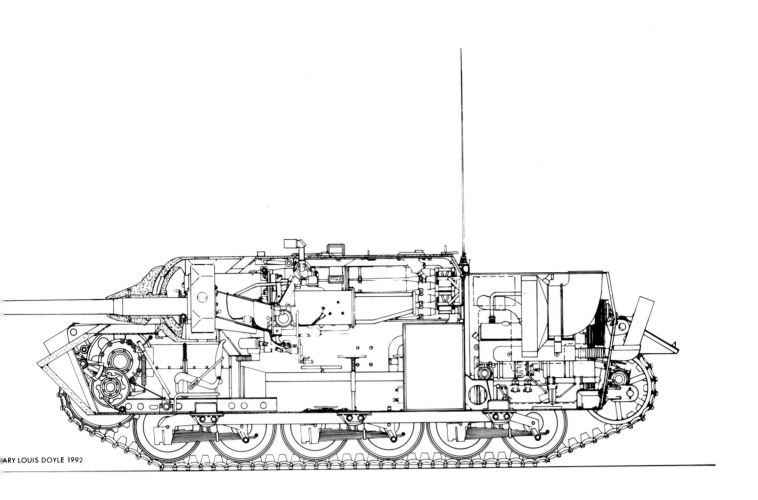

The new layout of the engine cooling of the Panzer IV lang (E) by the Porsche firm.

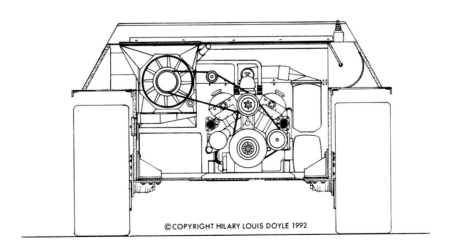

Four of these cooling units were built by the Skoda firm, which also tested them successfully until the end of August 1944. WaPrüf 6 requested an enlargement of the cooling surface and blower, in order to rule out loss of cooling by incoming dust. Skoda could only present a wooden dummy of the improved version, which fell victim to an air raid in October 1944.

Since all developmental work on the Panzer III/IV chassis was halted in November 1944, no test results are available.

Another design (Porsche Type 262) concerned improvements to the exhaust system on the Panzer III/IV chassis.

This project was taken up in February 1944 at the urging of WaPrüf 6. The exhaust temperatures of the Maybach HL 120 motor rose so high that the temperature in the engine compartment was 15% higher than the normal temperature. Porsche tried by remantling the exhaust pipes to create an additional sucking effect to cool the exhaust gases. Here, too, no test results exist.

Tracks

The tracks used until then were 400 mm wide, and thus had a too-high ground pressure. There was the possibility of building tracks with a width of 540 mm, which gave a better ground pressure. Because of the large number of pieces, tracks with cast and rolled parts were chosen.

Conclusions

The chassis of the light *Panzerjäger* and the Panzer IV tank had to be made identical. The following components were called for:

Powerplant:	Maybach HP 120 TRKM
Cooling system:	like Panzer IV (if a simpler system could not be made meanwhile)
Transmission:	ZF SSG 77 with Panzer III steering drive
Intermediate gear:	Panzer III (strengthened); if not strong enough, new development by Daimler-Benz
Steering brake:	Argus disc brakes. Testing these brakes should be extended at once, and carried out on at least five vehicles
Running gear:	Six-wheel running gear, Krupp suggestion, in line, with rubber-saving road wheels (660 mm diameter). Their suspension should be simplified if possible. - the working of the rounding of the crank arm in the mounting block should be dropped, except the sealing. - optional version of pressed-in road-wheel rods. Tests with this running gear to be done on a broad basis with five vehicles. - the Alkett firm was later called on to develop running gear with rubber-saving road wheels (660 mm diameter), with two wheels suspended on each truck.
Aprons:	Should be installed so they could be set vertically or angled.
Tracks:	540 mm wide, symmetrical, with one-tooth drive, if possible with one cast and one rolled link.
Leading-wheel	Leading wheels like Panzer IV, but larger adjusting for tracks with rolled links. The leading-wheel suspension was to be strengthened and simplified.
Suspension:	rolled links. Suspension should be strengthened and simplified.
Hull:	Panzer IV and *Panzerjäger* the same, with pointed bow. Fuel fillers not removable from below if necessary. Fire safety for the fighting compartment must be maintained.

Torque converter: Should be dropped if possible and be replaced by a cable that allows limited lateral weapon movement.

A conference at Krupp-Gruson in Magdeburg from March 15 to 17, 1944, also dealt with the 6-roller running gear for the "Uniform Vehicle III/IV." The test vehicle (chassis no. 84779) was back in Magdeburg on 15 March, and was essentially repaired as follows:

- The running-gear levers were fitted with wearing bushings,
- The cast journal bearing of the leading wheel was tested for wear.

Until the production of the new components, a short test drive with grease lubrication of the road wheels was carried out. They had lost oil because of too-great bearing play.

As already noted in this chapter, efforts were made for as broad a base as possible in testing the Uniform Running Gear III/IV. There were also five 8.8 cm Pak 43/1 on Panzer III/IV (Sf) chassis "Nashorn" without weapons (formerly "Hornisse," Sd.Kfz. 164). The situation in the spring of 1944 is documented as follows:

Hornisse I

Since production had already progressed too far, running-gear levers with pressed-in hardened bearing pin—without wearing bushings—with journal bearings had to be used.

The total weight of this Hornisse was to be brought to 30 tons because of the assault armor.

Hornisse III

Running-gear levers with pressed-in pins were installed, wearing bushings and cast alloy journal bearings.

Hornisse IV

Running-gear levers with pressed-in pins, wearing bushings and cast alloy journal bearings with various consistencies were installed.

Hornisse V

Bearing pins pressed into the running-gear lever, floating bushings. Pins without wearing bushing, hardened. This version could later be used without the floating bushings.

On March 22, 1944, the Army high Command gave the Krupp-Grusonwerk AG firm of Magdeburg-Buckau contract no. SS 4911/0006-3036 for the building and assembling of three Panzer IV on the Uniform Chassis. For this, three complete chassis (hulls) from Contract WaPrüf 6/Pz III, SS 4911/0006-3374/43 were delivered by the Alkett firm of Berlin-Borsigwalde.

For running-gear tests with the Panzer IV lang (E), self-propelled *Panzerjäger* mounts of the "Hornisse" type were called for.

On May 16, 1944, Krupp of Essen gave the OKH a set of drawings for the Light *Panzerjäger* III/IV with 7.5 cm *Panzerjägerkanone* (L/70).

On June 9, 1944, Krupp of Essen notified the Nibelungenwerk that the turret of the Uniform Chassis III/IV remained practically the same as that of the production series Model 10/BW (*Panzerkampfwagen* IV, Type J). Also on June 9, 1944, the OKH Chief of the Army and B.d.E. gave the Krupp firm of Essen War Contract SS 4912/0210 8808/44 for 1200 hulls and upper bodies for the *Panzerjäger* III/IV.

On June 28, 1944, the OKH changed the SS 4912/0210 - 8808/44 contract to Krupp of Essen as follows:

	Formerly	Now
1. Hulls for the *Panzerjäger* III/IV	1200	- -
2. Bodies for *Panzerjäger* III/IV	1200	- -
3. Hulls for *Panzerkampfwagen* III/IV	- -	610
4. Bodies for *Panzerkampfwagen* III/IV	- -	391

A message of July 18, 1944, stated: The production of the Panzer IV as a tank is also stopping at the Nibelungenwerk. All assault guns with the 7.5 cm Kanone L/70 were to be called "Panzer IV lang" at the Führer's orders. The assault gun on uniform chassis was to be designated **Panzer IV lang (E)**.

On September 12, 1944, the General of the Artillery reported to the Chef GenStdH on a visit to WaPrüf 6 in Kummerdorf, along with Colonel Holzhäuer and Lieutenant Colonel Kuebart on the development of the assault guns:

"The *Sturmgeschütz* Panzer IV lang (Uniform Chassis), which has been coming to the troops since about October 1944, will weigh 28 to 29 tons. A new air-cooled Diesel motor with some 350 HP is in development; for the time being, the Panzer IV lang (E) will be delivered with the Maybach HL 120 (265 HP)."

A message from the Krupp firm to WaPrüf 6 of September 23, 1944, gives information on the state of development shortly before it was broken off:

"Re: Running Gear - Uniform Chassis

On the basis of contracts no. SS 4911/0006 -3492/42 and -3493/43, the Gruson Works finished five rotary wheel-truck running gear with 660 diameter rubber-sprung road wheels. In June of this year, one set of the running gear (intended for Hornisse IV) was turned over to the Alkett firm of Berlin. The Alkett firm bestowed a contract for the new production of this running gear.

Meanwhile, all five running gear have been delivered, four of them to the Alkett firm.

We inquire now whether the contract given to the Alkett firm must still be filled. The parts will soon be finished, and we would like to hold them as replacements that we may need for the running gear now being tested. Otherwise, we would have to look for other replacements. We also ask for information as to what is to become of the hulls for Hornisse I, IV, and V."

Although at this point in time the further development of the Uniform Chassis III/IV attracted only meager interest, the plans for 1944 held fast strictly to this program.

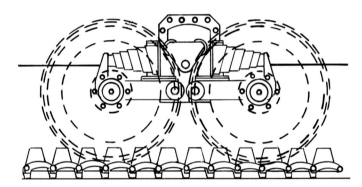

Alkett suggestion for the improvement of the suspension of panzer IV lang (E). Wheel trucks with horizontal truncated-cone springs and larger-diameter road wheels.

Wa A on June 7, 1944
<u>Az. 76 g 31 a Wa J Rü (WuG 6) VIIIa</u>
No. 2436/44 geh.
To
Reichsminister für Rüstung und Kriegsproduktion
z. Hd. Herrn HDL Saur.

Betr.: a) le.Pz.Jäg. 38 /t)
 b) Pz.Jäg. Vomag mit 7.5 cm Pjk L/70
In agreement with the H.A. Panzer, the following delivery plan is foreseen:

Type	May	June	July	Aug.	Sep.	Oct.	Nov.	Dec.
Le.Pz.Jg. 38 (t) 7.5cm Pjk L/48	50	100	200	300	400	500	600	700
le.Pz.Jg.IV 7.5 cm Pjk L/70	-	-	75	180	180	180	180	180
le.Pz.Jg.III/IV	-	-	-	-	-	-	10	50

This production means a high output of the le.Pz.Jg.38(t) with 7.5 cm Pjk L/48 to 1000 Pieces per month, to be reached in March 1945.

With the conversion of the former Assault Guns on the basis of Pz.Kpfwg. III and Pz.Kpfwg. IV on Uniform Chassis III/IV it is planned to bring the foreseen final capacity of 800 pieces per month as follows:

a)   600 pieces per month with 7.5 cm Pjk L/70
b)   125 pieces per month with 7.5 cm Pjk L/48
c)   75 pieces per month with 10.5 cm Stu.H. 42

The 200 units foreseen under b) and c) represent the share set by the Gen.St.d.H. for the Artillery assault gun units.

A Panzer Commission meeting on October 4, 1944, determined that the Tank Chassis III/IV should no longer be used in the future (for reasons of limiting armament concentration of the numbers of tank production types), and as of about the middle of 1945 only the following armored vehicles may be produced: Hetzer, 38 (t), Panther, and Tiger.

Development of Designations for the Panzer IV LANG (E)

Krupp full-size wooden model of the *Jagdpanzer* III/IV with 7.5 cm Pak L/71.

le.Pz.Jg. Wa Prüf 6	1/5/1944
Panzerjäger **III/IV** WuG 6	5/4/and 7/6/1944
Leichter Panzer *Jäger* III/IV mit 7.5 cm Pjk 42 (L/70)	
Krupp, Essen	5/16/1944
Le.Pz.Jg.III/IV mit 7.5 cm Pjk L/70	
GenStdH/General der Artillerie War Diary	6/7/1944
Sturmgeschütz **auf Einheitsfahrgestell III/IV**	
Führer's conference	7/6/1944

7.5 cm *Sturmgeschütz* III/IV L/70 WuG 6   7/14/1944
All *Sturmgeschütze* **mit 7.5 cm L/70** are, on the Führer's orders, designated
"Panzer IV lang." The *Sturmgeschütz* **auf Einheitsfahrgestell** is designated
"Panzer IV lang (E)   7/18/1944
Stu.Gesch.n.A. auf Einheitsfahrgestell
GenStdH/General der Artillerie War Diary 7/27/1944 **Panzer IV lang (E) m. 7.5 cm Pak 42 L/70**

"Overview of the Army's Armament State" Chef H Rüst u. BdE/Stab Rüst III   8/15 to 9/15/1944
le.Pz.Jg.III/IV Wa A Abn.   8/31/1944
Called by the troops: **Panzer IV lang**
Called officially: **Panzer IV lang Ausf.E1, E2 usw. (Einheitsfahrgestell)**
Chef GenStdH/Org.Abt./Gen.Insp.d.Pz.Tr.   9/11/1944
Sturmgeschütz **Pz.IV lang**
GenStdH/General der Artillerie War Diary   9/12/1944

View of the fighting compartment of the *Jagdpanzer* III/IV.

Appendix:

Jagdpanzer IV - Comparison

Comparison of Armor Thickness 60 and 80 mm
(60 mm until Vehicle 300, 80 mm as of Vehicle 301)

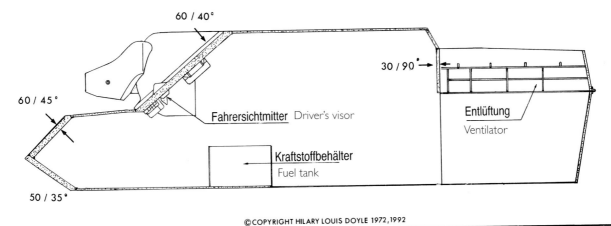

60 / 40°

60 / 45°

30 / 90°

Fahrersichtmitter Driver's visor

Entlüftung
Ventilator

Kraftstoffbehälter
Fuel tank

50 / 35°

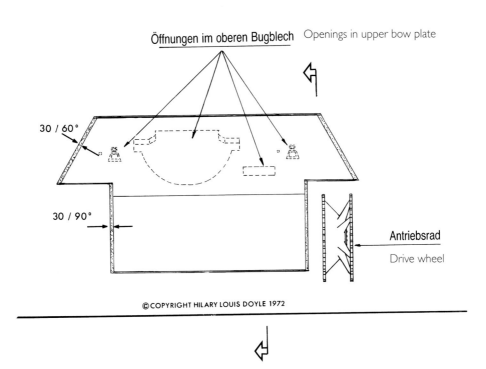

Öffnungen im oberen Bugblech Openings in upper bow plate

30 / 60°

30 / 90°

Antriebsrad
Drive wheel

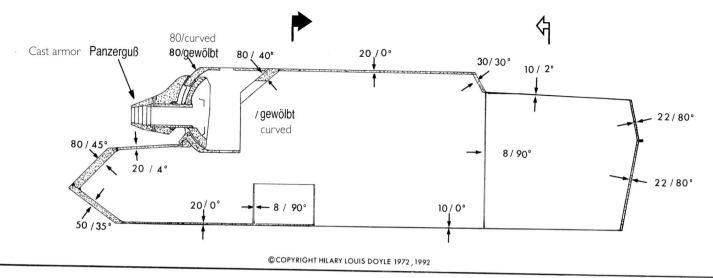

Cast armor **Panzerguß**
80/curved
80/gewölbt
80 / 40°
20 / 0°
30/30°
10 / 2°
22 / 80°
/ gewölbt
curved
22 / 80°
80 / 45°
20 / 4°
8 / 90°
20 / 0°
8 / 90°
10 / 0°
50 / 35°

©COPYRIGHT HILARY LOUIS DOYLE 1972, 1992

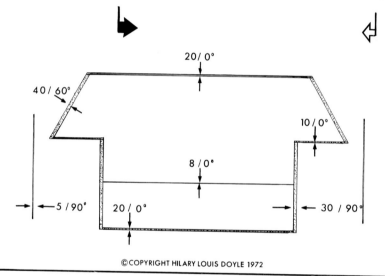

20/ 0°
40 / 60°
10 / 0°
8 / 0°
5 / 90°
20 / 0°
30 / 90°

©COPYRIGHT HILARY LOUIS DOYLE 1972

5 / 30°
22 / 78°
22 / 78°
10 / 0°

©COPYRIGHT HILARY LOUIS DOYLE 1972

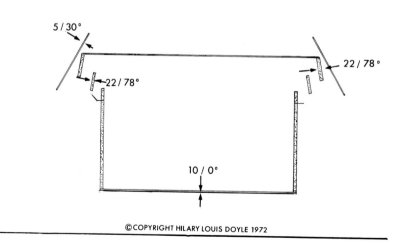

Panzerkleinzerstörer

Armored Small Destroyers

Since the beginning of 1942, various designs for small tank destroyers were worked out. The war diary of the General of the Artillery reports on January 7, 1944:

"Information on the state of development of various models (Hetzer,* Rutscher, explosive carriers with dummies, etc.) requested, in order to be able to offer advice on the development of a model."

In a meeting of the Army General Staff on February 28, 1944, Colonel Crohn (Wa.Prüf 6) reported that the production of the le.Pz.Jg. auf 38(t) was finished, and the projects noted above were halted.

* At this time, "Hetzer" was the evocative name for an unknown small *Panzerjäger* type.

The small armored destroyer "Rutscher" (Slider) of the BMW firm. The 1:1 scale wooden model shows the relative sizes of the vehicle and crew.

During a meeting of the Panzer Development Commission in Berlin W. 8, Pariser Platz 3, on January 23, 1945, General Thomale expressed, among other things, that small *Panzerjäger* projects were being taken up again because of the situation in the armored sector. In the east, great numbers of spare parts and heavy armored vehicles were being lost. Therefore he proposed:

- Maintenance efforts, spare-parts deliveries, and means of rescue
- In the framework of the shipped quantities of raw materials, making as many vehicles as possible. He welcomed the efforts in the realm of the *Jagdpanzer* 38(D). From this shortage of materials, the armored vehicle had at the moment become an armored defensive vehicle. But this was in no way attributable to an overly heavy cannon and large armor thicknesses. He portrayed how a great number of ideas and suggestions had dealt for two years with the so-called **Panzerkleinzerstörer**, the U-Boat of the dry land. At

the meeting, representatives of the *Luftwaffe* also took part, and wanted to have such vehicles for their paratroops.

Colonel Holzhäuer reported that since the spring of 1942 he had worked out some 20 designs, of which supply the BMW **Rutscher** and Weserhütte had become known.

The small tank destroyer (Panzer-*Abwehrfahrzeug*) conceived by the Bavarian Motor Works AG was powered by a production motor. The in-line six-cylinder motor that was used in the BMW Type 335, with a bore of 82 mm and a stroke of 110 mm—displacement 3485 cc—developed a power of 90 HP at 3500 rpm. The installation of the ZF AK 5-25 all-wheel drive was being considered. ZF had originally suggested to BMW the installation of the 6 EV20 electric drive, since a one-man crew was foreseen. But after a two-man vehicle was to have been built, the AK 5-25 drive was the better solution. Under the given circumstances, ZF did not have to stick to electric drive for this example. WaPrüf 6 promised on January 25, 1945, to check to see whether the use of the AK 5-25 drive would be possible.

The wooden model of the "Rutscher" made by MW, with built in recoilless double armament (designed circa 1943).

"Rutscher" project: transmission and steering gears.

"Rutscher" project (1944): steering gears.

"Rutscher" project: reduction gears.

"Rutscher" project" angular gear with hydraulic clutch.

For the "**Rutscher**" concept, BMW built a full-size wooden model as well as a 3550 mm long, about one meter high hull. BMW also provided the drive train, including the shifting and steering gears. Since nothing else had happened in the meantime, Büssing-NAG was to be given the order to build a two-man vehicle weighing three tons and using the air-cooled Tatra 4-cylinder, 90 HP motor.

Der *Generalinspekteur der*   **H.Qu.OKH, March 19, 1945**
Panzertruppen
Gr. Entwicklung No. 96/45 g/Kdos.

Memorandum
On a discussion about development questions with the chief of Staff of the Inspector-General of the Panzer Troops on March 14, 1945

Armored Small Destroyer

a) Vehicle of required characteristics, especially in weight of ca. 3.5 tons, absolutely requires new components (motor, gearbox, running gear, etc.), the development and production of which will allow series production in at least one and a half to two years. For reasons of time, therefore it is not to be followed up.

b) Only a vehicle that would weigh some 7-10 tons when finished can be formed out of presently available major-series components and be put into production quickly.

c) The 600-700-meter main combat distance effective range of the PK 8 H 63 can only be installed if the vehicle is appropriately small (thus, a weight of 3.5—at most 5 tons), lively, and nimble. A 7- to 9-ton vehicle that, to be effective on heavily armored, heavily equipped opponents, must approach to 700 meters, will be rejected by the Inspector-General of the Armored Troops under present-day conditions as tactically worthless.

d) The "Armored Small Destroyer" project must therefore, under consideration of the present developmental and production situation, be allowed to stop. The test pieces (3.5 ton Weserhütte or 7.5 ton Daimler-Benz) will be finished and delivered for technical-tactical testing. Besides the PWK weapon, installability of the 7.5 cm Cannon L/48 is also to be considered because of the mastering of combat distances over 700 meters.

On March 25, 1945, Lieutenant Stollberg reported that, for the planned small destroyer, an FAK 45 gearbox and 150 HP Saurer Diesel motor were planned. The vehicle was to be developed by Daimler-Benz. The gearbox was located next to the driver. Thus, the shift cover with lever had to be attached so that the gearbox could be transverse, whereby the shift-lever design was carried out just as with the AK 5-80 gearbox (with the same parts). On the other hand, only one arrangement of the waves over each other was possible with a flat-lying shift lever with vertical movement—similar to the deliveries from Sweden. ZF recommended an arrangement as in the AK 5-80 gearbox; this allowed the lowest construction. The driveshaft had to have double bearings, and be fitted with a connecting flange for the cardan drive. ZF also suggested the installation of a clutch brake. As a gearbox ratio, a complete range of 8.13 was suggested. Such a one was to be created only after a long advance period. ZF therefore suggested the available ratio of 7.5. A delivery could be promised immediately, since this gearbox was in production. It was the ZF FAK 45 P 10 experimental gearbox for small destroyers. The planned armament was the *Panzerabwehr-Wurfkanone* PWK 8 H 63 (designation as of November 1944, formerly Device 5-0864) developed by Rheinmetall.

This **high-** and **low-pressure** weapon worked by the following principle: The ignition of the shell took place in a **high-pressure space** in the cartridge through powder burning under high pressure. The entry of the gases into the **low-pressure space** throttled a jet plate. The shot separated from the cartridge as soon as a certain pressure had occurred in the low-pressure space, and left the barrel with even pressure.

This new principle allowed a significant reduction in weight and work since, among other things, a thin-walled, smooth gun barrel could be used.

The first firing tests with 8 cm Wgr. 5071 and 8 cm W Hl Gr. 4462 took place on November 8, 1944, in Hillersleben South. Muzzle velocities of some 550 m/sec, and shot ranges of about 2000 meters were achieved.

The plans foresaw the development of the weapon both singly and in several examples.

Klöckner-Humboldt-Deutz suggested a compromise solution that was based on available components. This vehicle weighed some five tons, and was thus literally no longer usable by airborne troops. Five such vehicles were to be assembled quickly, but series production never took place.

Developmental Vehicles

The newly established Research Group in the Army Weapons Office was supposed to create a new series of armored vehicles.

Colonel Holzhäuer reported on the progress with these vehicles in the meeting of the Panzer Development Commission on January 23, 1945. The Types E 10 and E 25, for which ZF was contracted to supply a rear electric transmission, should be developed as far as possible. The development of the E 10, though, had been halted in the interim, since the *Jagdpanzer* 38 was technically superior to it. To complete the presently published materials, there were now details about the E 10 at hand. The design by Klöckner-Humboldt-Deutz, Magirus Works, foresaw an armored model, protected in front by 60 mm, and 30 mm thick plates above and below.

Jagdpanzer E 10
(after original documents of the Klöckner-Humboldt-Deutz AG, Magirus Works).

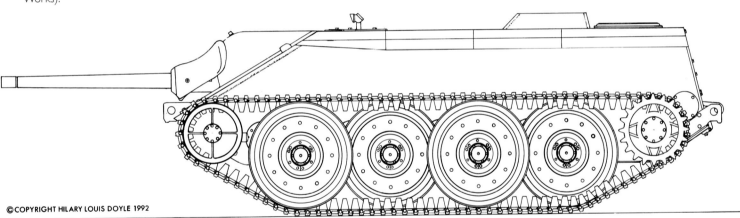

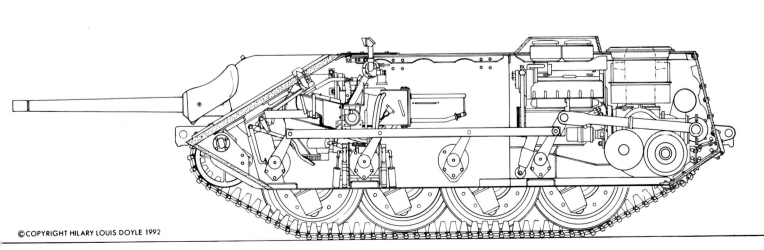

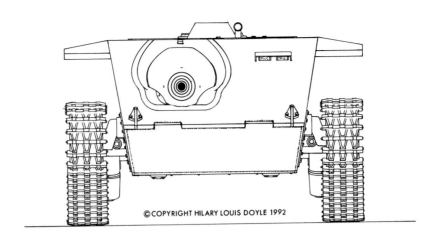

©COPYRIGHT HILARY LOUIS DOYLE 1992

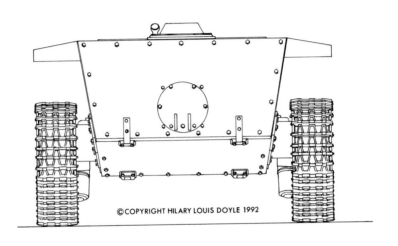

©COPYRIGHT HILARY LOUIS DOYLE 1992

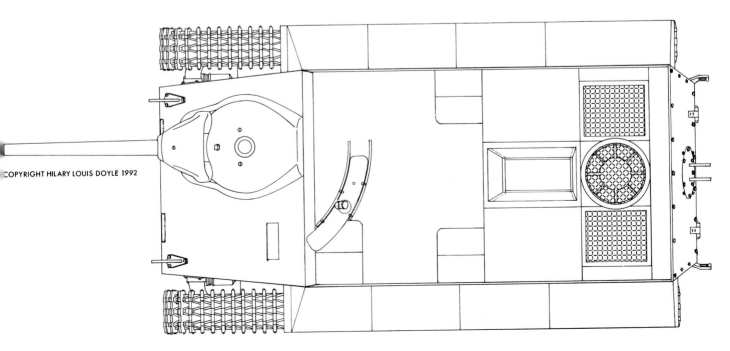

COPYRIGHT HILARY LOUIS DOYLE 1992

The rest of the armor plates were 20 mm thick. The entire vehicle could be lowered mechanically from 1760 mm to 1400 mm in height. The ground clearance was 400 mm. The planned motor was a Maybach HL 100 gasoline type producing 400 HP. An alternative was an air-cooled 350 HP Argus motor.

The J. M. Voith firm in Heidenheim/Brenz had developed a power aggregate on a hydrodynamic basis for the fast, light types of the E series. The vehicles weighed between 15 and 25 tons, reached a top speed of 65 to 70 km/h, and produced 350 to 400 horsepower. The shifting and steering gears, including the cooling system, formed a compact organic unit that was wholly housed in the rear of the vehicle. This aggregate could be removed as a whole for repairs or inspection after the mechanics had removed the engine compartment cover.

The gearbox consisted of two torque converters, one firmly attached, the other connected with the driveshaft in two-speed shifting via a changing step while in action.

To connect the steering with the gearbox, the two lateral planetary-drive intermediate gears worked via a zero shaft together with the reversible intermediate wheel. In straight-line driving the zero shaft stayed motionless, but on curves it was driven according to the turning direction in one or the other direction alternatively via hydraulically regulated clutches.

Through the filling regulation of these two hydraulic steering clutches, the range from the smallest to greatest radius could be taken without doing it in stages.

From this development, which was done under Voith's code word "Arta," parts of several prototypes were found in production when the war ended.

The three-man crew of Fahrzeug E 10 had a 7.5 cm Oak 39 L/48 available as the primary weapon. The further development of the Type E 25, planned for the 7.5 cm *Panzerjägerkanone* L/70, was already prepared for. According to Colonel Holzhäuer, the hulls for the prototypes were already at Kattowitz. Mr. Von Heydekampf was of the opinion that the E 25 vehicle could fill a hole in the weight class between the 38 (D) and Panther in the distant future. The Argus firm in Karlsruhe was responsible for this development.

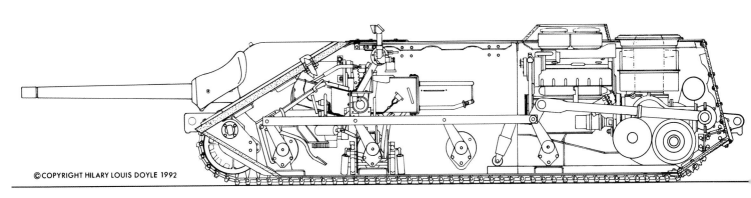

Jagdpanzer E 10, lowered.

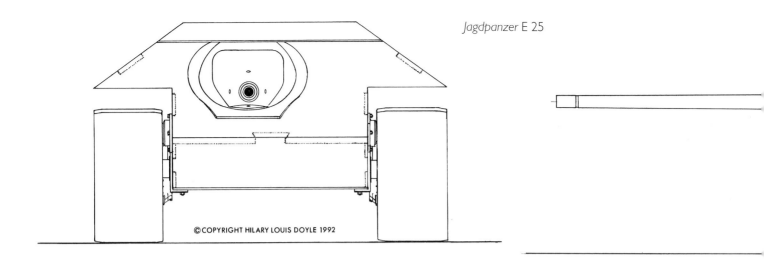

Jagdpanzer E 25

©COPYRIGHT HILARY LOUIS DOYLE 1992

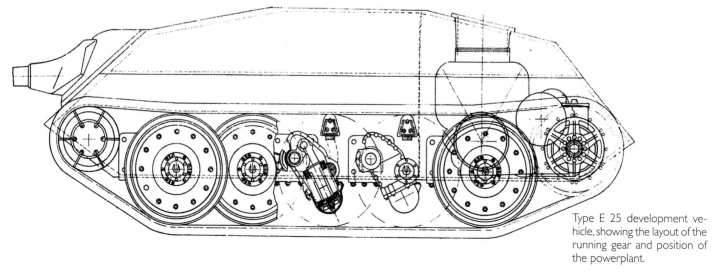

Type E 25 development vehicle, showing the layout of the running gear and position of the powerplant.

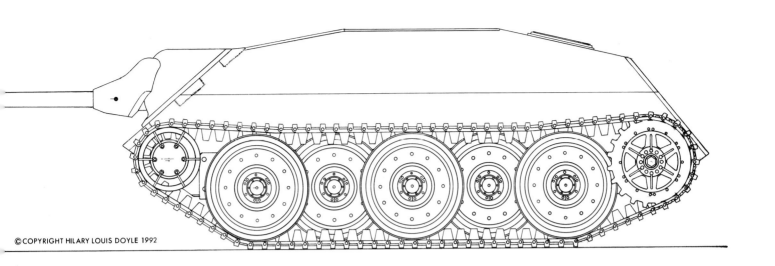

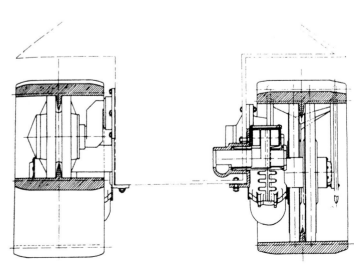

Type E 25: details of the independent wheel suspension with spring elements.

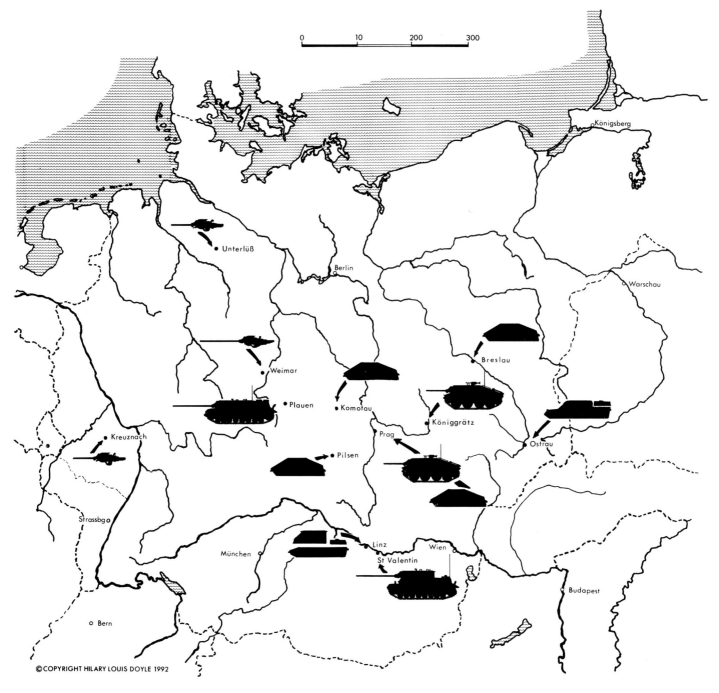

Locations of the manufacturers.

Supplying Firms

Components	Firm	Production code
7.5 cm Pak 39 L 48	Seitz-Werke, Bad Kreuznach	dcm
	Rheinmetall-Borsig, Unterlüss	nyv
Jagdpanzer 38 hull	Poldihütte, Komotau	dpm
	BMM, Prague	epa
	Linke-Hofmann, Breslau	dpj
	Skodawerke, Königgratz	bxb
Jagdpanzer 38	Böhmisch-Mährische Maschinenfabrik, Prague	epa
	Skodawerke, Königgratz	ews
Jagdpanzer IV hull	Witkowitzer Bergbau und Eisenhütten	
	Gewerkschaft, Ostrau	bzs
Jagdpanzer IV	Vomag, Plauen	ajk
7.5 cm Stuk 42 L 70	Gustloff-Werke, Weimar	bcd
	Rheinmetall-Borsig, Unterlüss	nyv
	Skodawerke, Pilsen	bxb
Hull		
Panzer IV/70 (A)	Eisenwerke Oberdonau, Linz	kmy
Panzer IV/70 (A)	Nibelungen-Werk, St. Valentin	hhv

Technical Data

Manufacturer	BMM, Skoda	BMM, Skoda, Alkett	Vomag
Type	*Jagdpanzer* 38	*Jagdpanzer* 38 D	*Jagdpanzer* IV Type F
Years built	1944-45	1945	1944
Data source	D 652/63 of	Various	D 653/39 of
	11/1/1944	9/15/1944	
Motor	Praga epa AC 2800	Tatra TD 103 P*	Maybach HL120TRM
Cylinders	6 in-line	V-12	V-12
Bore x stroke mm	110 x 136	110 x 130	105 x 115
Displacement cc	7754	14820	11867
Compression ratio	6.5:1	17:1	6.5:1
RPM	2800	2250	2600
Maximum HP	160	207	265
Valve type	drop	drop	drop
Crankshaft bearings	7 journal	7 journal	7 roller
Carb./Fuel Injec.	2 Solex 46 FNVP	Bosch PE 12	2 Solex 40 IFF
Firing order	1-5-3-6-2-4	1-8-5-10-3-7-6-11-	1-12-5-8-3-10-6-7-
		12-12-12-12	2-11-4-9
Starter	Bosch BPD 3/12	Bosch BPD 6/24	Bosch BNG 4/24
Generator	Bosch GQLN 300/	Bosch RKCK 300/	Bosch GTLN 600-/
	12-900	12-1400	12-1500
Batteries/Volts/Ah	1/0/100	2/12/150	4/12/105
Fuel feed	pumps	pumps	pumps
Cooling	liquid	air, blower	liquid
Clutch	dry disc	dry disc	dry disc
Transmission	Praga-Wilson	ZF AK 5-80	ZF SSG 77
Gears fwd./reverse	5/1	5/1	6/1
Drive wheels	tracks, front	tracks, front	tracks, front
Intermediate ratio	7.33 (as of 11/44, 8)	5.58	3.23
Top speed km/h	40	40	40
Range km	road 180, off 130	510	road 210, off 130
Steering type	planetary	planetary	clutch
Turning circle m	4.54	4.54	5.92
Suspension	leaf composite	leaf composite	° longitudinal
Brake effect	mechanical	mechanical	mechanical
Brake type	band	band	band
Brakes work on	drive	drive	drive
Running gear	road wheels & jacks	road wheels & jacks	road & jack wheels
Vehicle track mm	2123	2183	2480
Track type	Kgs 350/140	-	Kgs 61/400/120
Ground length	mm	3020	2700 3590
Links per track	96	-	99
Track width mm	350	420	400
Ground clearance mm	380	-	400
Long/wide/high mm	6270/2630/2100	6850/2600/2400	6960/3170/1960
Ground pressure	0.78 kg/cm²	-	0.85 kg/cm²
Fighting weight kg	16000	16700	24000
Load limit kg	1500	2000	2000
Crew	4	4	4 (command veh. 5)
Fuel consumption liters/100 km	road 180, off 250	off road 76	road 220, off 360
Fuel capacity liters	320	390	470 (3 tanks)
Armor: hull front mm	60	60	80
hull side mm	20	20	30
hull rear mm	20	10	20
body front mm	60	60	60
body side mm	20	20	40
body rear mm	20	20	30
Primary weapon	7.5 cm Pak 39 L/48	7.5 cm Pak 42 L/70	7/5 cm Pak 39 L/48
Other weapons	1 MG 34	2 MG 42	1 MG 42
Ammunition rounds	40	62	79
Upgrade degrees	25	30	30
Climbing mm	650	650	600
Wading mm	1100	900	1000
Spanning mm	1500	1800	2200
Notes	called "Hetzer"	* 0 series with Tatra	(Sd.Kfz. 162)
	V8 Diesel engine		

Manufacturer	Vomag	Alkett	Alkett/Krupp/MIAG
Type	Panzer IV/70 (V)	Panzer IV/70 (A)	*Jagdpanzer* III/IV
Years Built	1944-45	1944-45	1944-45
Data source	various	various	various
Motor	Maybach HL 120 TRM		
Cylinders	V-12	V-12	V-12
Bore x stroke mm	105 x 15	105 x 115	105 x 115
Displacement cc	11867	11867	11867
Compression ratio	6/5:1	6.5:1	6.5:1
RPM	2600	2600	2600
Maximum HP	265	265	265
Valve type	drop	drop	drop
Crankshaft bearings	7 roller	7 roller	7 roller
Carb./Fuel inject.	2 Solex 40 IFF	2 Solex 40 IFF	2 Solex 40 IFF
Firing order	1-12-5-8-3-10-6-7-2-11-4-9		
Starter	Bosch BNG 4/24	Bosch BNG 4/24	Bosch BNG 4/24
Generator	Bosch GTLN 600 12-1500		
Batteries/Volts/Ah	4/12/105	4/12/105	4/12/105
Fuel feed	pumps	pumps	pumps
Cooling	liquid	liquid	liquid
Clutch	dry plate	dry plate	dry plate
Transmission	ZF SSG 77	ZF SSG 77	ZF SSG 77
Speeds fwd./reverse	6/1	6/1	6/1
Drive wheels	tracks, front	tracks, front	tracks, front
Intermediate ratio	3.23	3.23	3.23
Top speed km/h	35	38	38
Range km	road 210, off 130	road 320, off 210	road 320, off 210
Steering type	clutch	clutch	clutch
Turning circle m	5.92	5.92	5.92
Suspension	° longitudinal	° longitudinal	horiz. trunc. cone
Brake effect	mechanical	mechanical	mechanical
Brake type	band	band	Argus disc
Brakes work on	drive	drive	drive
Running gear	road & jack wheels	road & jack wheels	road & jack wheels
Vehicle track mm	2480	2480	2500
Track type	Kgs 61/400/120	Kgs 61/400/120	Kgs 61/400/120
Ground length mm	3590	3590	3980
Links per track	99	99	99
Track width mm	400	400	400
Ground clearance mm	400	400	400
Long/wide/high mm	8600/3170/1960	8960/3330/2236	8600/3060/2080
Ground pressure	0.86 kg/cm^2	0/86 kg/cm^2	0.86 kg/cm^2
Fighting weight kg	25800	28000	28000
Load limit kg	2000	2000	2000
Crew	4 (command, 5)	4 (command, 5)	4 (command, 5)
Fuel consumption Liters/100 km	road 220, off 360	road 20, off 360	road 20, off 360
Fuel capacity l	iters	470 (3 tanks)	680 450
Armor hull front mm	80	80	60
hull side mm	30	30	30
hull rear mm	20	20	20
body front mm	80	80	80
body side mm	40	40	40
body rear mm	20	20	20
Primary weapon	7.5 cm Pak 42 L/70	7/5 cm Pak 42 L/70	7/5 cm Pak 42 L/70
Other weapons	1 MG 42	1 MG 42	1 MG 42
Ammunition rounds	55	60	60
Upgrade degrees	30	30	30
Climbing mm	600	600	600
Wading mm	1000	1000	1000
Spanning mm	2200	2200	2200
Notes	(Sd.Kfz. 162/1)	interim model	only prototypes

THE SPIELBERGER GERMAN ARMOR AND MILITARY VEHICLE SERIES

Heavy Jagdpanzer

Development • Production • Operations

Walter J. Spielberger

Hilary L. Doyle

Thomas L. Jentz

A SCHIFFER MILITARY HISTORY BOOK

HEAVY JAGDPANZER
Development • Production • Operations
Spielberger, Doyle, & Jentz

The complementary volume to the *Light Jagdpanzer*, this volume in the series presents the *Wehrmacht's* heavy tank destroyers (Ferdinand, *Jagdpanther*, and *Jagdtiger*) with previously unpublished photos and action reports.

Size: 81/2"x11" • over 260 b/w photos, line schemes • 200 pp.
ISBN: 978-0-7643-2625-7 • hard cover • $49.95